Y0-BDL-714

Robert Robinson
Chemist Extraordinary

Robert Robinson
Chemist Extraordinary

TREVOR I. WILLIAMS

CLARENDON PRESS · OXFORD
1990

Oxford University Press, Walton Street, Oxford OX2 6DP
Oxford New York Toronto
Delhi Bombay Calcutta Madras Karachi
Petaling Jaya Singapore Hong Kong Tokyo
Nairobi Dar es Salaam Cape Town
Melbourne Auckland
and associated companies in
Berlin Ibadan

Oxford is a trade mark of Oxford University Press

Published in the United States
by Oxford University Press, New York

British Library Cataloguing in Publication Data
Williams, Trevor I. (Trevor Illtyd)
Robert Robinson—Chemist extraordinary.
I. Organic chemistry. Robinson, Sir Robert, 1886–
I. Title
547'.0092'4
ISBN 0–19–858180–7

Library of Congress Cataloging in Publication Data
Williams, Trevor Illtyd.
Robert Robinson, chemist extraordinary
Trevor I. Williams.
1. Robinson, Robert, Sir, 1886–1975. 2. Chemists—England—Biography. I. Title.
QD22.R63W54 1990 540'.92—dc19 [B] 89–22123 CIP
ISBN 0–19–858180–7

Set by Hope Services (Abingdon) Ltd.
Printed in Great Britain by
Bookcraft (Bath) Ltd., Midsomer Norton, Avon

Preface

For those seeking information on the life and work of Robert Robinson there already exists a considerable volume of published material. Apart from the numerous obituary notices which appeared in the general and scientific press shortly after his death in 1975, there are several more substantial sources. The most important of these, in order of their publication, are his own unfinished autobiography, *Memoirs of a Minor Prophet* (1976); the long tribute by Lord Todd and Sir John Cornforth in the *Biographical Memoirs of Fellows of the Royal Society* (1976); Lord Todd's entry for the *Dictionary of National Biography (1971–80)*; and the proceedings (published *in extenso* in *Natural Product Reports (1987)*) of a special meeting of the Historical Group of the Royal Society of Chemistry held at Warwick University in 1986—as part of the Society's Annual Congress—to mark the centenary of his birth. Yet even the summation of these varied contributions falls short of an overall portrayal of Robinson's versatile genius and complex personality. In particular, reflecting the nature of the publications in which they appeared, they lay more emphasis on his professional achievements than on his personal life.

Because of this gap in the literature I was responsive to a request from Dr Marion Way, Robinson's daughter, to write an extended biography of her father, on the recommendation of Lord Todd and Sir John Cornforth. There were, additionally, personal reasons urging my acceptance. I was at that time putting the finishing touches to my biography of Howard Florey (1984) and there was—as will appear—some community of interest between the two men. I came to be writing Florey's biography because during the war I had been a member of his research term in the Sir William Dunn School of Pathology, and the reason why I found myself there was because Robinson had recommended me. Thereafter, I came to know him well, and had a great regard and affection for him: over many years he showed me much personal kindness. To write his official biography was thus, in a sense, to repay a debt.

While this biography thus to some limited extent reflects my personal knowledge and experience it could not have been written without a great deal of help from many sources. Considering the breadth of his interests, and the enthusiasms with which he pursued them over an exceptionally long life, the amount of archival material that has survived is disappointingly small. This is partly because, although an energetic correspondent, most of his letters were written in longhand, so that no copies exist. Such personal papers as were recovered from Lady (Stearn) Robinson were catalogued by Mrs Jeannine Alton and Miss Julia Latham-Jackson of the Contemporary Scientific Archives Centre, then in Oxford, and subsequently deposited in the library of the Royal Society in London. I am much indebted to the Society's then Librarian, Mr N. H. Robinson, for aiding my researches on this material.

Many family papers and other memorabilia have been made freely available to me through the kindness of Mr Robert Bradbury Robinson, the present Chairman of Robinson and Sons of Chesterfield. Mr George Walsh, Robinson's nephew, now resident in Cape Town, has also provided much helpful information, including some annotated letters from his uncle. I have also had helpful correspondence with Robinson's niece, Mrs Shirley Sampson. On the other side of the family, Lady (Stearn) Robinson's daughter, Mrs Stephanie Cottrell—now of Madison, Connecticut—has been most helpful in providing me with background information about her mother.

I must record my thanks also to others for their personal, rather than professional, recollections. Among these are Mr Helmut Meyer, Lady (Stearn) Robinson's literary agent in New York; Mr Tom Corbett, co-author with her of *The Dreamer's Dictionary*; Mr Roald Dahl, the Robinsons' neighbour at Great Missenden; and Miss Stella Corridon, Robinson's secretary during part of his time with Shell.

Naturally, however, I have relied most heavily on those who, directly or indirectly, were associated with Robinson in a professional capacity, and the list of those whose help I must acknowledge is a long one. I am particularly indebted to Dr Graham Holland, of the University of Sydney, for the immense amount of trouble he took to provide information concerning Robinson's appointment as professor of chemistry there more than seventy-five years ago: without this, that chapter would have been regrettably thin. Similarly, I must thank Dr G. N. Burkhardt for providing much valuable background information on Robinson's years at Manchester. Dr John Shorter, of the University of Hull, made available to me his specialist knowledge of the history of the electronic theory of valency. At an early stage in the project Sir Edward Abraham very kindly let me have copies of some of his correspondence with Robinson and others on the chemistry of penicillin. Professor David Ginsburg, of the Technion in Haifa, provided me with photocopies of many letters which he exchanged with Robinson over a period of some twenty years. Learning of my project Professor Maurice Stacey, of the University of Birmingham, wrote to me at length about his correspondence and contacts with Robinson over a long period. A particular word of thanks is due to Sir Ewart Jones, whose friendship I have enjoyed over many years. As Robinson's successor as Waynflete Professor of Chemistry at Oxford, and with an exceptional knowledge of the chemical world, he was particularly well placed to advise me, especially about the fortunes of the Dyson Perrins Laboratory during the last years of Robinson's stewardship. Again in Oxford, I have been much helped by Dr R. H. Jaeger, who worked with Robinson in Oxford and moved with him to the Shell laboratory at Egham when he retired in 1955. I have enjoyed talking about Robinson and his days at Oxford with Dr Leslie Sutton, who was elected a Fellow of Magdalen in 1936, only a few years after Robinson's own arrival. Over a period of some twenty years Robinson was actively involved with Mr Robert Maxwell and Pergamon Press; I am grateful to Mrs Robert Maxwell for the trouble she has taken to make a great deal of archival material available to me.

As the book developed I sought information on many aspects of Robinson's life, and much was also offered spontaneously. In this context I must thank the following, among others: Dr R. M. Acheson, Professor Wilson Baker, Dr B. K. Blount, Professor R. N. Chakravarti, Mrs J. B. Cottis, Dr G. I. Fray, Mr and Mrs G. C. C. Gell, Mr H. K. Hartley, Mr F. Hilton, Dr Alexander King, Dr E. D. Morgan, Mr David Nealy, Dr A. Pollard, Dr G. S. Pope, Dr R. Schoental, Professor Charles Shoppee, Dr N. Stein, and Dr A. F. Thomas.

I have also to thank officers of the Universities of Sydney, Liverpool, St Andrews, Manchester, London (University College), and Oxford for providing information relevant to the various appointments held by Robinson. Ms Nehama Shalom, of the

Weizmann Institute, very kindly went through the voluminous correspondence of Chaim Weizmann kept there, and sent me photocopies of letters relating to Robinson.

Apart from making heavy demands on time over a long period, works such as this involve much incidental expenditure. It could not have been undertaken without the help of grants from the Leverhulme Trust, the E.P.A. Cephalosporin Fund, and the Royal Society. These I gratefully acknowledge.

The full text of this work has been read and approved by Dr Marion Way and, at her request, by Dr Paul Kent, who has had a long association with the schools of chemistry and biochemistry in Oxford; I appreciate his interest and comments. Relevant sections have been read by Sir Edward Abraham, Dr Graham Holland, Sir Ewart Jones, Mrs Robert Maxwell, and Dr John Shorter. I am grateful to them and to all those who have helped me, but must stress that the views expressed, and any errors of commission or omission, are my responsibility.

I end on a personal note. In a sense, writing a biography is like having a house guest on an extended visit. It is proper, therefore, that I should thank my wife for welcoming into our home the shades of Howard Florey and Robert Robinson in succession over a period of more than ten years: for a time, both of them together. She knew them both, and our desultory conversations about the activities and idiosyncrasies of our guests were very helpful in sorting out my ideas as the book progressed. It became even more a family affair when my daughter Clare lent a hand with the typing (and retyping).

Of my biography of Florey one friend, whose opinion I greatly valued, was kind enough to remark that he thought Florey would have liked it: I hope very much that Robinson might have felt the same about this one.

Oxford T.I.W.
1989

Contents

The Perkin family tree

PART I

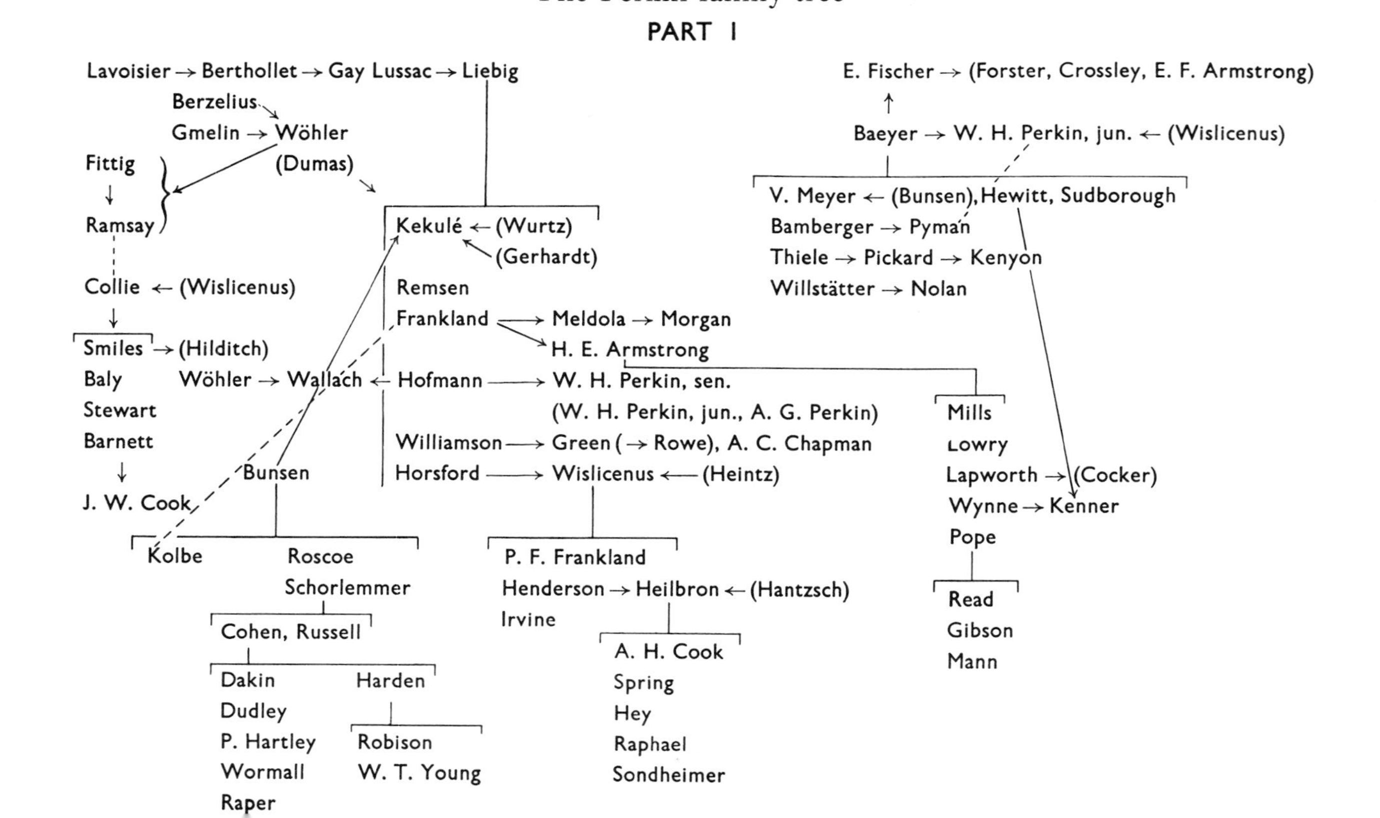

PART II

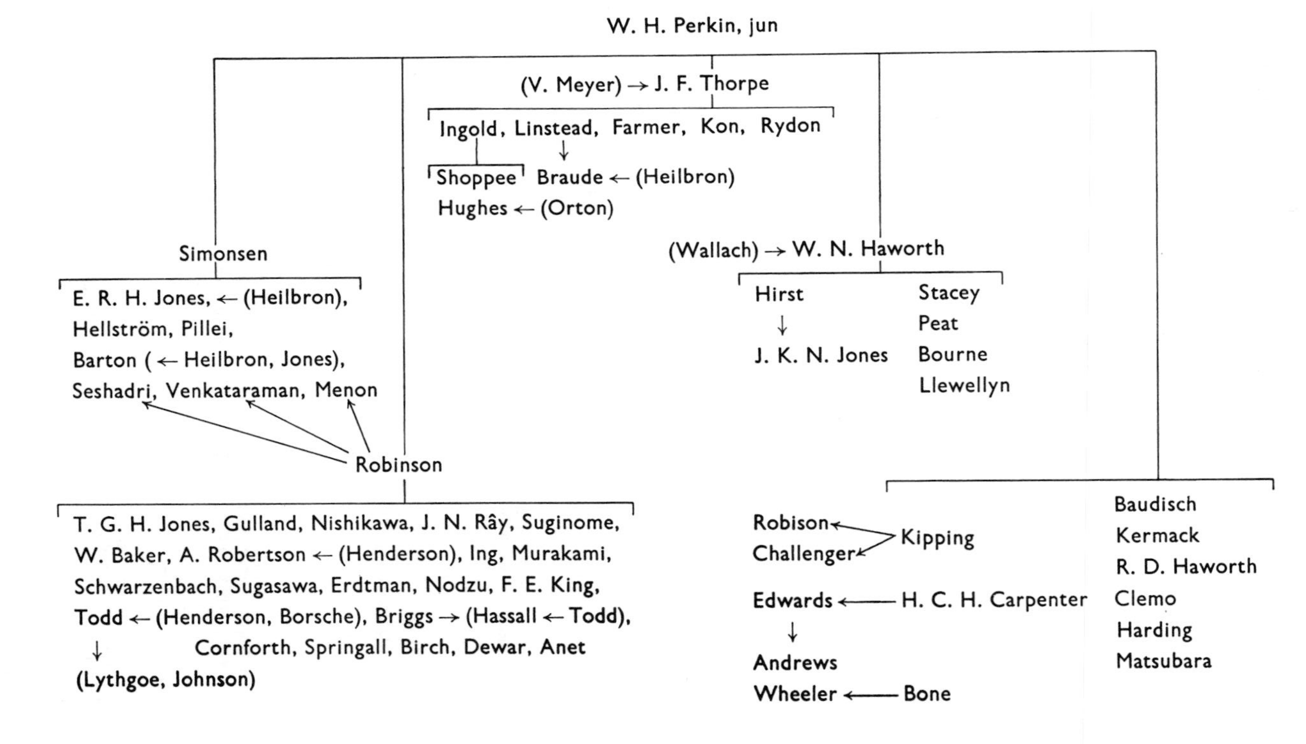

W. H. Perkin, jun
(V. Meyer) → J. F. Thorpe
Ingold, Linstead, Farmer, Kon, Rydon
Shoppee
Braude ← (Heilbron)
Hughes ← (Orton)
(Wallach) → W. N. Haworth
Hirst
J. K. N. Jones
Stacey
Peat
Bourne
Llewellyn
Simonsen
E. R. H. Jones, ← (Heilbron),
Hellström, Pillei,
Barton (← Heilbron, Jones),
Seshadri, Venkataraman, Menon
Robinson
T. G. H. Jones, Gulland, Nishikawa, J. N. Rây, Suginome,
W. Baker, A. Robertson ← (Henderson), Ing, Murakami,
Schwarzenbach, Sugasawa, Erdtman, Nodzu, F. E. King,
Todd ← (Henderson, Borsche), Briggs → (Hassall ← Todd),
Cornforth, Springall, Birch, Dewar, Anet
(Lythgoe, Johnson)
Robison
Challenger
Kipping
Edwards ← H. C. H. Carpenter
Andrews
Wheeler ← Bone
Baudisch
Kermack
R. D. Haworth
Clemo
Harding
Matsubara

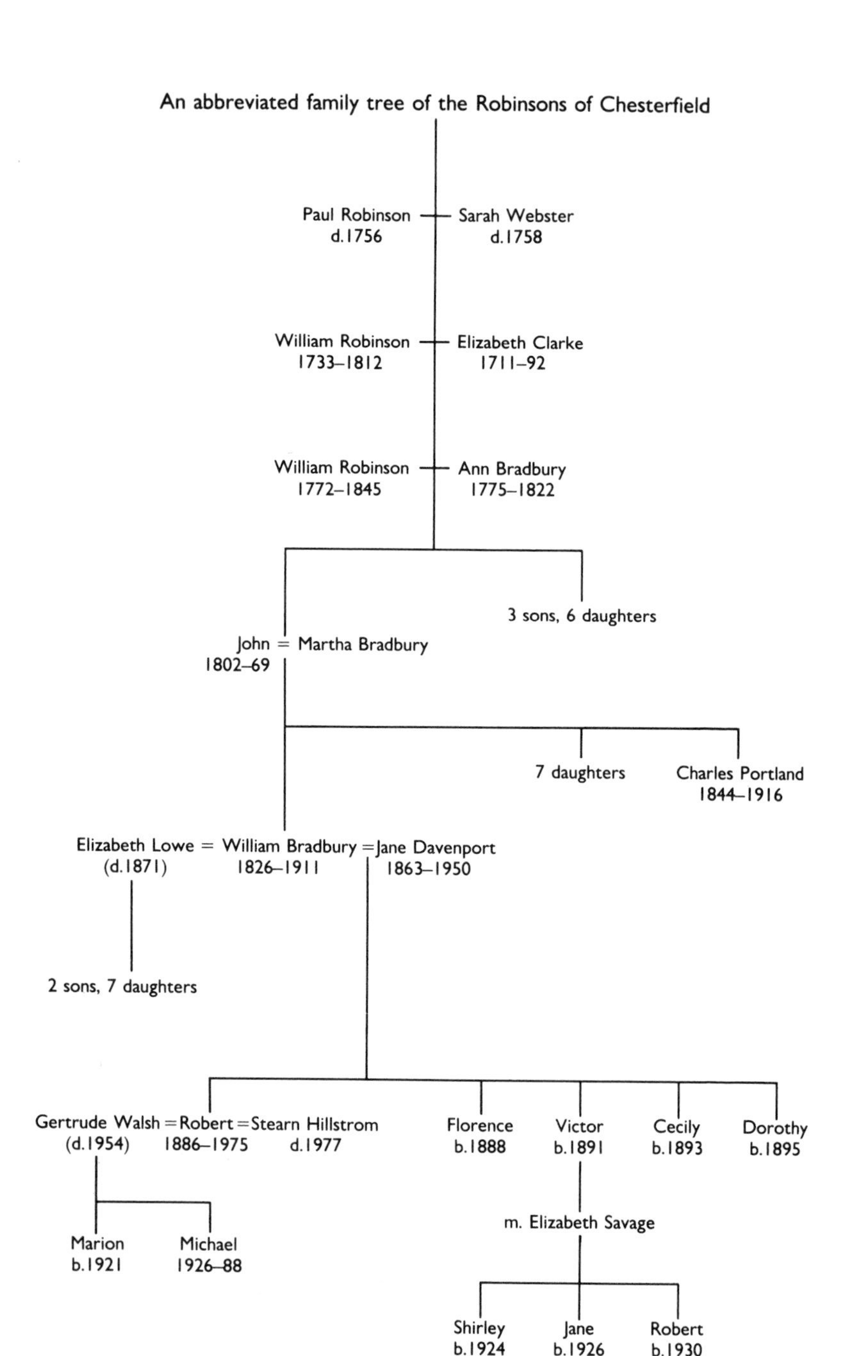
An abbreviated family tree of the Robinsons of Chesterfield
Paul Robinson
d.1756
Sarah Webster
d.1758
William Robinson
1733–1812
Elizabeth Clarke
1711–92
William Robinson
1772–1845
Ann Bradbury
1775–1822
3 sons, 6 daughters
John = Martha Bradbury
1802–69
7 daughters
Charles Portland
1844–1916
Elizabeth Lowe = William Bradbury = Jane Davenport
(d.1871)
1826–1911
1863–1950
2 sons, 7 daughters
Gertrude Walsh = Robert = Stearn Hillstrom
(d.1954)
1886–1975
d.1977
Florence
b.1888
Victor
b.1891
Cecily
b.1893
Dorothy
b.1895
Marion
b.1921
Michael
1926–88
m. Elizabeth Savage
Shirley
b.1924
Jane
b.1926
Robert
b.1930

I
The Robinsons of Chesterfield

The relative influence of nature and nurture on character and ability can never be precisely defined, but it cannot be doubted that family background and upbringing are significant factors in shaping the characters and careers of individuals. In the case of Robert Robinson we have a quite exceptional opportunity of exploring family background and relationships, for one member of the family, Philip Robinson, devoted much time in the early 1930s to compiling an exhaustive family history—covering nine generations—of the Robinsons of Chesterfield. This appeared in 1937 as a handsome 320-page volume[1], for private circulation within the family; to this was added a 136-page supplementary volume[2] in 1961. That this mine of information ever saw the light of day was purely a matter of chance. In 1935, Philip—a member of the still thriving family business, to which we will come later—found his normally active business life interrupted by the Depression. To occupy himself in this period of unwonted leisure, he turned to some notes for a family history collected by his father, Charles Portland Robinson: this so captured his imagination that, until business revived in 1937, he devoted much of his time to its preparation. Later, in retirement, he wrote the supplement to correct and extend the original work.

The Robinsons are an old-established Derbyshire family, with collateral branches initially identified particularly with the neighbouring towns of Chesterfield and Bolsover. A Robert Robinson lived in Chesterfield in the sixteenth century, for the parish register for 14 April 1565 records the baptism of his son Thomas: thereafter the name occurs frequently in the district in the context of birth, marriage, or death. The Bolsover register records Robinsons from 1698, and that of Brampton—10 miles to the north—from 1752. However, the first generation of Robinsons to be specifically identified in the History were living in Chesterfield in the early part of the eighteenth century. There Paul Robinson was married in 1722 to Sarah Webster, by whom he had eight children. He seems, however, to have worked briefly in the 1740s with a brother who had a pottery at Bolsover. In 1750, however, he returned to Chesterfield with his son William, and set up his own business as a potter in Spa Lane.

Chesterfield at that time was a flourishing market town, with a population of about 8000, whose first Charter dated from 1215. Its parish church dates from the fourteenth century and was, and remains, a local landmark because of its twisted spire, the result of the warping of insufficiently seasoned timber. Long the centre of an agricultural community, it was by the early eighteenth century experiencing the beginning of the manufacturing enterprise that later was to develop as the Industrial Revolution. There were good local supplies of coal and iron-ore, and the canal to the Trent, completed in 1776, gave improved transport. Paul Robinson, as we have noted, was a potter, and over the years members of the family were much involved with business enterprises of one sort or another. Paul's son William (1733–1812) also pursued the potters' trade in Chesterfield, after gaining experience with his uncle in Bolsover, and took up the manufacture of clay pipes, including the long-stemmed variety known as a 'churchwarden'. The making of pipes remained a local industry until 1839. By the end of the eighteenth century, however, the new fashion for snuff greatly reduced the demand for pipes; and in 1791 William sent his only child—also William, but universally known as Billy (1772–1845)—away to Manchester to learn the drapery trade. Billy did well, and by 1794 had his own drapers' shop in Chesterfield; he also manufactured gloves, and ran his father's pipe-making business. By 1835 he was trading as 'William Robinson and Son, Linen and Woolen Drapers'. Thus we see the beginning of the association with the textile industry which was to be an integral part of Robert Robinson's family background.

In 1796 a new family relationship was created. In that year Billy married Ann Bradbury, daughter of Josiah Bradbury of Chapel-en-le-Frith: the alliance was probably initiated through a member of the Bradbury family then living in Chesterfield. By this marriage he had ten children: his eldest son, Josiah Bradbury Robinson (1798–1877), succeeded him in the business, and was active in it until his death in 1877. Josiah was evidently a man of standing in the community, for in 1848 he was an executor for George Stephenson, the railway pioneer, who spent the last year of his life at Tapton

Hall, and is buried in Trinity Church, Chesterfield. Stephenson's gold tie-ring, given to him in recognition of this, remains a family treasure. Robert's half-brother, William Bradbury Robinson (1863–1924) married into the Stephenson family in 1897.

In 1825 the Bradbury family connection was strengthened when Billy's third son John Bradbury Robinson (1802–69) married his cousin Martha Bradbury. In this marriage we can possibly see some direct influence on our Robert, for the family's predominating interest in trade and industry was reinforced by one in science. In 1818, at the age of 15, John was apprenticed for five years to Joseph Claughton, in business in Chesterfield as a chemist and druggist; subsequently he set up independently in the same line of business at Belper, but shortly transferred to Chesterfield. His health was poor, but his mind alert and active. His daily regimen was strict: 'Rise every morning at the latest 6½ . . . Let all your work be done by 7½ in the evening'. His diary shows that this regular 13-hour stint was devoted not only to classical studies such as Latin, philosophy, and theology, but also to mathematics, mineralogy, botany, and archaeology, as well as pastimes such as gardening and drawing.

In 1839 he decisively altered his way of life, selling his chemists' business and buying a near-bankrupt firm known as the Druggists' Pill Box business, thus laying the foundation of the firm that in due course became Robinson and Sons, Ltd. His health was poor and the business did not flourish. Suddenly, and apparently without a word of warning to anybody, he abandoned it all and set sail for America in 1841, leaving his wife Martha to manage the business and their young family as best she could. The detailed diary he kept shows that while in America he manufactured, and traded in, pill-boxes, but in a desultory way: his main interest was clearly in observing the rights and customs of an exciting new country. He returned home in 1842, apparently in rather better health, and in 1854 formed a business relationship with his son William Bradbury Robinson (1826–1911), Robert's father.

If we are in search of inherited traits we can possibly see in John Bradbury Robinson a pale reflection of his grandson Robert. He had a lively and erratic mind, which he was as ready to devote to the problems of business—though not with conspicuous success—as to the acquiring of scientific knowledge. Robert was born after his death in 1869, but from his *Memoirs*[3] he clearly regarded him with approval.

It is, of course, largely to parents and siblings that one must normally look for the strongest childhood influence, but before turning our attention to Robert's immediate family, we must for a moment go back on our tracks and follow another important thread running through the Robinson family.

As we have seen, their general background was in manufacture and trade on a modest scale; for the most part, they made a comfortable living and had a standing in local society, but had no wider aspirations. Another common bond between them lay in religion; like many others of similar background they often held—sometimes very actively—staunch Nonconformist views, though not always of the same persuasion. Following the Act of Uniformity in 1662 the parish priest of Chesterfield, a Presbyterian, and his curate, a Congregationalist, were—like some 2000 ministers up and down the country—expelled, but some of the congregation remained loyal to them. For a time they had no regular place of worship, but following the Toleration Act of 1689 a joint Congregational/Presbyterian meeting house was built in 1692 in what is now Ellar Yard. Over the years, however, the Presbyterians veered strongly towards Unitarianism; and in 1778 they built the Blue Meeting House in Froggett's Yard (so called because it was built of local blue slate), and later, in 1822, Soresby Street Congregational Church.

As Paul Robinson was married in the parish church of Chesterfield, and all eight of his children were baptized there over the period 1733–44, we can presume that he adhered to the Church of England; but his son William (1733–1812) became a Presbyterian. His son Billy, too, was ardent in the Nonconformist cause: the History records that:

> He delighted in hearing the famous preachers of the day when in London and attended services morning, afternoon, and evening.

He was a trustee of the Old Blue Meeting House and—with his sons Josiah and John—took an active part in the foundation of the new chapel in Soresby Street. As he served also as a Director of the Chesterfield and North Derbyshire Bank he may be held to illustrate the traditional Mancunian belief in 'One God, One Devil, and Twenty Shillings in the Pound'.

And so, at the end of a long line of solid, Nonconformist business men—broken only by the truant husband John Bradbury Robinson—we come to Robert's father, William Bradbury Robinson, born at

Belper in 1826. He, too, followed in the local Nonconformist movement, first at Soresby Street and later at the new Brampton Congregational Chapel, and entered the family business. However, this traditional path was not followed without some resistance on his part. After local education, he found work in the pill-box factory irksome, and sought to break away and follow a seafaring life. Accordingly he set sail for America in the SS *Portland* on 28 September 1844—a date of some significance, for it was the day on which his youngest brother was born: the latter was consequently christened Charles Portland (1844–1916). He returned early in the following year, but set sail again in 1848, with high hopes, not fulfilled, of establishing his own factory in New York. Returning home again, he established his own pill-box works in Brampton alongside that of his father, with whom he eventually joined forces in 1854: even then, however, he kept his own business going independently until 1860.

William Bradbury Robinson's special line was the manufacture of ointment boxes from very thin strips of willow. These were very widely used by the pharmacists of the day; but a growing shortage of willow in England, and increasing competition abroad, seriously affected the business. Some diversification was, therefore, necessary, and it was decided to embark on the manufacture of surgical dressings. How this decision was reached we can only surmise, but various reasons suggest themselves. The pill-box business provided a link with the medical profession through the chemists and druggists, and there was a natural interest in textiles in a town no great distance from the great textile towns of Lancashire and Yorkshire: moreover, William Bradbury's grand-father, Billy, had himself at one time engaged in the textile trade. More immediately, the Crimean War (1854–6) created a new demand for dressings; and, while this would obviously be relatively short-lived, the possibilities for general surgery were much enhanced by the advent of anaesthetics, and, later, Lister's technique of antisepsis. Be this as it may, 3 July 1854 saw Robert's father setting off for Derby in search of a second-hand lint frame advertised in the local paper. This was duly installed, and the first lint was made in the following January. However, this was a clumsy treadle machine; and John Bradbury quickly applied his mechanical bent to the design of appropriate power-looms. From 1854 the business was known as John B. Robinson & Son—now Robinson and Sons Ltd—and an early price-list describes it as a 'Lint, Cartonnage, Paper and Willow

Box Factory'. Over the years it developed as a highly successful family business, but Robert never played any active part in it, though in later life he became a member of the Board. Its immediate interest is that it was an important part of his family background in his formative years.

Today, in the western world, the death of a child is sufficiently unusual to be a matter for particular grief, but a century and more ago things were very different. Although the number of children was often very large, the high rate of child mortality kept families small. The case of William Bradbury Robinson (1800–76), a hosier, was not untypical. All his ten children by his first marriage predeceased him, and six died in infancy. The second William Bradbury Robinson, Robert's father, was more fortunate: by his first marriage to Elizabeth Lowe he had ten children, of whom three died in infancy: of the survivors one was a son and the others six daughters. After Elizabeth's death in 1871, aged 43, he remarried in 1885. His second wife was Jane Davenport, a member of an old Cheshire family connected with the silk industry centred at Macclesfield, who worked in the office of the box factory at Wheatbridge: he was then 59 and his bride only 21. He lost no time in starting a new family, and Robert was born less than a year later. His birthplace was Rufford Farm; but when he was three years old they moved to Field House, New Brampton. He was followed by Florence (1888), Victor (1891), Cecily (1893), and Dorothy (1895).

Why, after being a widower for some 14 years, and having already a considerable family, William Bradbury decided to remarry, and chose a bride so very much younger than himself can, now be only a matter for speculation. It is not, however, an uncommon occurrence. But this combination of high survival rate among the children and the remarriage late in life put Robert in an unusual family situation. The difference in age between his oldest and his youngest sisters was 44 years, and some of the grandchildren of the first marriage were older than the children of the second. To unravel the resulting complicated family relationships in a way intelligible, let alone interesting, to the general reader is virtually impossible, especially as many of the girls acquired familial names different from their baptismal ones: thus Mary was commonly known as Polly and Martha as Arna. By the time Robert was born almost all the members of the first family had left home, and came to Field House only as occasional visitors, especially for family reunions at

Christmas. The only exception was Martha, the eldest, who as a spinster lived in a separate extension of Field House, and was, therefore, in much closer contact.

While details are merely confusing, some general comment is not without relevance. With the exception of Robert, the Robinson family followed a fairly predictable course. William Bradbury II, the only son of the first family to survive infancy, entered the business straight from school. Victor served with distinction during the First World War, and afterwards took a keen interest in the Territorial Movement. Florence, the eldest daughter, served as a nursing orderly during the war at the Endell Street Military Hospital in London, and then returned home to work in the welfare department of the business. She took a keen interest in local politics, and eventually (1946) became Mayor of Chesterfield, the fifth member of the family to hold this office. Cecily, too, took up nursing during the war, and was a V.A.D. in France; afterwards she carried on this career, and after some years at St Thomas' in London joined the Overseas Dominions Nursing Service and worked in East Africa and Hong Kong, where she was interned after it fell to the Japanese. Dorothy, the youngest, in 1920 married Alex Bell, who was for many years in charge of the Robinson office and warehouse in London.

This is a catalogue of solid achievement, a closely knit local family, parochial in the best sense of the word, with its interest centred upon Chesterfield and the family business there. They married into local families of their own standing in trade and industry, and were not given to venturing far afield. Although instigated and compiled by Philip Robinson, the remarkable History—specifically for private circulation amongst members of the Robinson family—could never have been prepared without the wholehearted co-operation of relatives, many quite distant. This in itself emphasizes the strong sense of family unity. There is nothing in the family history to give an inkling that there might suddenly appear a man of rare genius, who would break right away from the family tradition, achieve world-wide recognition as a brilliant scientist, and gain the highest distinctions his country could confer on him. There is, of course, nothing exceptional in this, for genius, scientific and otherwise, often springs from very humble origins: Newton's father was described as 'wild, extravagant, and weak', and Humphry Davy's father was a wood-turner and Faraday's a blacksmith. Pasteur was the son of a tanner. On the other hand, there are families where a high degree of talent appears regularly, almost as a matter of course. Such, for example, are the Cecils, the Rothschilds, the Wedgwoods, the Darwins, and the Huxleys. The source of Robert's genius must remain a mystery; and there is now no possibility of seeing whether it could blossom again in later generations.

In later life Robert had very decided views on organized religion; and the mixed influences to which he was subjected as a child may possibly have some bearing on this. As we have seen, the family had a very strong Nonconformist tradition; and his father was initially a member of the Soresby Street chapel, before being admitted to Brampton Congregational Church on 24 December 1874. He had thus been of this persuasion for half a century; but his second marriage, to the young Jane Davenport, was solemnized at Holy Trinity Church, Blackpool. We must suppose, therefore, that she was then a member of the Church of England, and that—to please his young bride and her family—William Bradbury agreed to his marriage taking place in a church whose very name was the antithesis of the unitarian view. Jane, for her part, seems to have been equally accommodating, for in due course she joined the Brampton Congregational Church, and it was there that she was buried in 1950, after being nearly forty years a widow.

In later life, as we shall see, Robert showed antipathy towards organized religion; but in his *Memoirs* he freely acknowledges that the religious atmosphere of his father's household decisively and beneficially influenced his life. Not, it is to be feared, in strengthening his faith: but in setting him on the path in life in which he was to be so remarkably successful. Both his father and his uncle, Charles Portland Robinson, were active in the affairs of the Congregational church, and its ministers were frequent visitors to Field House. Robert, early a free thinker, was not greatly impressed by what he overheard of their theological discussions, but eagerly accepted their invitations to play chess, a game for which he had a natural aptitude, and which gave him much pleasure until the end of his long life: even when afflicted by blindness in later years his remarkable memory enabled him to dictate his moves. The Robinson family being what it was, it had been assumed as a matter of course that Robert would go into the business immediately he left school; but the ministers—more perceptive than his father of changing educational needs—urged the desirability of a university education, something previously unknown, and probably never even considered, within the family. It was undoubtedly their strong and unexpected advocacy which set him on an academic career.

In those days, university entrance requirements in Britain were less formal than they are today, and a correspondingly less specialized education was acceptable as a preliminary. The Robinsons had no tradition of sending sons away to any of the great public schools, and for the most part they were educated locally. Typically, Robert's father had gone to a small school in Chesterfield run by a Mr J. Atkinson, but he chose a rather different pattern for his son. Robert went first to a kindergarten school run by a Miss Walton, and then briefly to Chesterfield Grammar School. There he might well have stayed, like other Robinsons before him; but at the age of twelve his father sent him as a boarder to Fulneck School at Pudsey, Greenside, half-way between Bradford and Leeds. This was a departure from custom, in that it was run by the Moravian Church, whose foundation in 1457 is particularly identified with the teaching of John Huss, who had been burnt as a heretic in 1415. Its members were dispersed under their bishop Comenius, a famous educational reformer who had been minister at Fulneck in Czechoslovakia in the seventeenth century; but the Church was refounded in 1722, and began a missionary movement that had considerable success. They were a strict community, recognizing only the authority of the Scriptures; but they were organized on presbyterian lines, and thus compatible with Congregationalists. Even so, the choice is unexpected: according to Robert the main reason was that Albert Bingham, a commercial traveller for the business for whom William Bradbury had a particular respect, had been at another Moravian school, Ockbrook, and recommended Fulneck strongly. Additionally, two of Robert's cousins had done well at a Moravian school in the Rhineland. He himself, in his *Memoirs*, says that the standard of teaching was remarkably high; and this is borne out by the fact that his younger brother Victor was also sent there in due course: briefly, they overlapped.

At home, Robert very much enjoyed the life of the countryside, and what he called 'a rather boisterous childhood'. Family discipline was light—perhaps the influence of a young mother—and there was little restriction provided no harm was done to people or property. In the winter there was skating and tobogganing—but above all there was climbing. A favourite climb on a steep outcrop called Castle Naze, above the little village of Combes where his great-grandmother Ann Bradbury had been born more than a century earlier. Climbing, like chess, became a lifelong passion, and he always recalled his debt to Morton Clayton, who had married his half-sister Lynie, for later introducing him to Alpine climbing.

Although the discipline at Fulneck was by his own account rather rough, the transition from Chesterfield seems to have been easy. Boarding was no particular problem, for he had already been a weekly boarder at Chesterfield Grammar School; and the surroundings were all he could have desired. The school was situated in the country on a hill above a deep valley, facing the picturesque village of Tong, and had good playing-fields. It was co-educational, with separate wings for girls and boys, separated by the chapel. Along the front was a long terrace, on which impromptu games of cricket could be played in the lunch break.

As schooldays go, the years at Fulneck seem to have passed agreeably enough; but although the notion of going on to university before entering the business had been agreed in principle, something quite specific had to be decided upon. Robert's father's wish was that he should study chemistry; and this was doubtless a carefully considered choice. At the time of which we are speaking there was a close association between the textile and the chemical industries, which had been established for more than a century. What more natural, therefore, than that if the son of a textile manufacturer wanted to enlarge his knowledge and experience he should turn to chemistry. Moreover, his father had already experienced the consequences of lack of chemical knowledge; his attempt to build a bleach works on the basis of an article in *Chambers Encyclopaedia* had been a notable failure. Robert, however, had different views. In his last year at School he had had individual tuition in mathematics from J. H. Blandford, who had taken the mathematical tripos at Cambridge. This was so much to his liking that he wanted to make this his subject at university. Apart from this, he had a leaning towards engineering. This may well have been acquired from his father, who delighted in mechanical devices and had his own well-equipped workshop. Robert's mechanical aptitude was demonstrated during a school holiday, when his father offered him a gold sovereign to design an automatic lint-cutting machine. This he did so successfully that the machine ran well for several years. The episode evidently meant much to him, for he never spent the sovereign, and it was eventually made into a brooch for his wife. In his *Memoirs* he states that 'if I had not taken to chemistry I think I would have been pleased to continue with machines'.

However, these might-have-been aspirations were recorded some seventy years later, with the nostalgia of old age, and should perhaps not be taken too seriously; they are akin to Einstein's declaration that had he not been a theoretical physicist he would have liked to have been a plumber. In later life, as we shall see, Robert's formulation of the electronic theory of organic reactions suffered from his reluctance to develop it mathematically, and he took no particular interest in mechanical devices. In any event, his father's wishes prevailed, and in his last year at Fulneck he took the matriculation examination of the Joint Board of the Universities of Manchester, Liverpool and Leeds and duly passed—somewhat to his own surprise, because he was conscious of his weakness in certain subjects. In 1902 he was admitted to Victoria University Manchester to study chemistry.

2

Students days at Manchester

Although the academic world was largely *terra incognita* so far as the business-oriented Chesterfield household was concerned, the decision that Robert should read chemistry at a university more or less dictated that he should do so at Manchester. Apart from its geographical convenience, located no more than forty miles from Chesterfield, this would be the natural inclination of such a strongly Nonconformist family. On the evidence of his later talent, he could no doubt have gained a place at Oxford or Cambridge had he been so minded; and after centuries of total classical ascendency the teaching of chemistry there was well established. At Oxford, the University Commission of 1854 had recommended the creation of university chairs in science. The Waynflete Chair in chemistry—which Robert was himself to hold in due course—had been founded in 1865, and at the turn of the century was occupied by William Odling. He was a distinguished chemist—Fellow of the Royal Society and sometime President of the Chemical Society—and his colleagues there included such other able men as N. V. Sidgwick and J. E. Marsh.

But north country Nonconformists did not share the traditional veneration of Anglicans for the University of Oxford, from which they had been firmly barred until some thirty years previously. Not until the University Tests Act of 1871 could students enter either Oxford or Cambridge without a religious declaration. If we may believe one of many anecdotes told of Benjamin Jowett, the formidable Master of Balliol, the admission test had by then become something of a formality. It is said that one young man who had come up to take his place in the College informed the Master that he had looked around the world and could find no evidence for the existence of God. 'Well' snapped Jowett 'you had better find some evidence by nine o'clock tomorrow morning, or I shall not admit you.' For those whose faith was weak, putting their name to the required form of words doubtless meant little, but in staunch Nonconformist circles it was quite unacceptable, and removal of the barrier did not immediately remove resentment at its earlier existence. It was against this background that a university had been founded in Manchester. There John Owens, in partnership with George Faulkner, had built up a large fortune as a textile merchant and railway speculator. Poor health led him to live the life of a recluse; but he took a keen interest in education, and had a particular detestation of all religious tests. Towards the end of his life he proposed making Faulkner his heir; but the latter protested that he already had sufficient, and generously proposed that Owen should instead leave his fortune—ultimately £96 000—for the endowment of a new university.

Thus was founded, in 1851, Owens College, Manchester, subject 'to the immutable rule and condition that the students, professors, teachers, and other officers and persons connected with the said institution shall not be required to make any declaration as to, or submit to any test whatsoever of their religious opinions.' How far such considerations would have weighed with Robert himself even at that stage in his career must be a matter for surmise, but they would certainly have weighed considerably with his father and other members of the family. Apart from this, an excellent school of chemistry had developed at Manchester in the half century that had elapsed since the foundation of the College. By a charter of 1880 it had become—together with University College, Liverpool and Yorkshire College, Leeds—a constituent member of the Victoria University, Manchester. Very shortly after Robert's admission this federal organization was terminated, and Manchester, Liverpool, and Leeds each gained independent university status. The new University inherited a strong local chemical tradition. In 1781 the Manchester Literary and Philosophical Society had been founded, and of this John Dalton—internationally famous for his concept that the atoms of different elements are distinguished by having different weights—was a member for nearly 50 years, until his death in 1844. Another very distinguished member was James Prescott Joule, famous for formulating the mechanical theory of heat. At Owens College the chemical laboratory had been presided over by (Sir) Henry Roscoe, a brilliant teacher and publicist who during his thirty years there (1857–87) made Manchester into indisputably the leading chemistry school in Britain. His assistant was Carl Schorlemmer, who in 1874 was made the first, and for many years the only, professor of organic chemistry in the country, a reminder of how new was the science which Robert was to make so particularly his own. Roscoe and Schorlemmer's *Treatise on Chemistry* was published in 1877, and for half a century, in successive editions, was a standard work.

In 1876 Roscoe had been succeeded by H. B. Dixon, whose education was unusual by modern standards, but less so by those of late nineteenth century Britain. From Westminster School he went to Oxford to read classics, but did poorly at this, and transferred to natural science, gaining a first class honours degree in 1875. However, he did not abandon the classics, and in later life was given to composing poems in Greek. At Oxford, he trained as a research chemist under the exacting tutelage of A. G. Vernon Harcourt; and in the year that he went to Manchester was elected Fellow of the Royal Society. Apart from his stimulating lectures on non-metallic elements, he must also have appealed to Robert as an alpinist: he climbed the Matterhorn three times without a guide and without a rope. Such, too, was W. H. Perkin Jr, who in 1892 had succeeded Schorlemmer as Professor of Organic Chemistry. Robinson's mountaineering enthusiasm must surely have been encouraged by its prevalence in the department. Perkin was a capable chemist in his own right, and the son of Sir William Perkin, who in 1856 had revolutionized the chemical industry with the discovery of mauveine, the first synthetic dyestuff. Outside the laboratory his great passion was music, and he was himself an exceptionally talented pianist.

The department contained several other considerable and stimulating characters. Among them was W. A. Bone, whom Robinson recalled as a man flamboyant in both dress and language, 'with a Mephistophelean air'; he had a keen interest in the technology of coal-gas, and in 1905 was appointed first Livesey Professor of Coal Gas and Fuel Industries at Leeds, and was elected Fellow of the Royal Society; subsequently he became Professor of Chemical Technology at Imperial College, London. Thus the department as a whole presented a stimulating combination of academic excellence and recognition of practical issues. No less important, many of the staff had, as we have seen, a remarkable variety of extra-mural interests: the Manchester of those days was a great cultural as well as industrial centre.

To Robinson, as a new student only seventeen years of age, these senior men were at first only names—people to be critically assessed at a distance on the strength of their lectures and laboratory teaching. Only later did some emerge as friends and colleagues, often lifelong. Initially, new friendships were made largely within the ranks of fellow students of his own age and here, too, he was fortunate. Through the Fulneck connection, he lived at first in the local Moravian College at Fairfield near Buxton, whose Principal was L. G. Hassé. His son, Ronald, with whom Robert had been at school, was reading mathematics at Manchester: he had a distinguished academic career, eventually becoming Professor of Mathematics at Bristol. Another lodger at the College was W. T. Waugh, studying history, who ended his career as Professor of History in McGill University, Montreal. Among student friends in the chemistry department was Peter Sandiford, who also eventually emigrated to Canada: tiring of chemistry, he took an active interest in education, and became Professor of Educational Psychology at Toronto. Robert recalls him as an ardent socialist, who practised what he preached during a holiday they took together in Germany. To understand how the poor lived Waugh insisted on travelling everywhere 4th class, which in those days meant an open carriage with no seats: as Robert ruefully recalled 'you took a suitcase and sat on it if you could.' He was not the only radical: another chemistry student was Rhona Robinson, a militant suffragette. Another friend was C. J. T. Cronshaw, who eventually went on to a distinguished career in ICI.

The cricket which he had enjoyed at Fulneck, where he had been Captain of the second eleven, led to another new friendship at Manchester. There he recognized among his fellow students the Captain of the second eleven of Silcoates School: both were primarily bowlers and in one match each had dismissed the other for a duck. This was Ashley Dukes who devoted much more time to the theatre than he did to chemistry; ultimately he abandoned this and made a name for himself as an author and dramatist: his Polish wife, Marie Rambert, founded the Ballet Rambert in London in the 1930s. Robert recalls that Dukes once introduced him to Bernard Shaw 'but the great man was far from scintillating on that occasion and, so far as I remember, said nothing of any interest or importance.'

With Ashley Dukes and other friends he enjoyed walking in the hills; but on winter evenings he found plenty of other diversions. In particular, he enjoyed the Music Hall, then in its heyday, and the occasional circus, and the pantomime at Christmas. As a populous and wealthy centre, with a number of excellent theatres, Manchester could attract such performers as Sybil Thorndyke and Lewis

Casson, Gracie Fields and Harry Lauder. The Gilbert and Sullivan Opera Company were also regular visitors and he greatly enjoyed their performances.

In his *Memoirs* he recalls his involvement in a mind-reading act at the Hippodrome by a couple known as the Zancigs. The husband would take articles at random from the audience and ask his blindfolded wife on the stage to identify them. Mischievously, Robert gave him a slip of paper with a chemical formula of brazilin scribbled on it. He was nonplussed when a very fair description of it came back from the stage, and much relieved to learn, years later, that it was all done by a complicated code.

Vacations, of course, were spent largely at home, especially for the great family reunions at Christmas. There, as ever, walking in the hills was a regular pleasure. An added attraction was a small laboratory which his father—doubtless anxious to further his son's chemical career as best he could—had built as an extra storey on an outhouse. Robert expressed his pleasure: 'I am glad Mr Rhodes has started on that room, and I shall certainly make no small use of it, and Victor, too, will use it, of course'. At this stage in his life his chemical interests were catholic, and he investigated the gaseous products resulting from the decomposition of air by a six-inch spark produced by an induction coil: these could be drawn off for examination with a spectroscope. To the layman such an intense and noisy discharge can be alarming, and an oblique reference in the *Memoirs* indicates that this particular line of research soon had to be curtailed. However, he was already displaying interest in the organic chemistry which was to be his lifelong work, and prepared terebic acid by a synthetic process then only recently (1900) invented by the French chemist Victor Grignard. No doubt this was prompted by Perkin's use of this technique for the synthesis of terpenes. This had an interesting sequel, when he mentioned it to John Simonsen, a fellow student of chemistry at Manchester. Simonsen had graduated in 1904, and was subsequently appointed the first Schunck Research Fellow in the University: he, too, developed an interest in natural-product chemistry, and the two became lifelong friends. In 1908 they went together to Dublin to attend the annual meeting of the British Association. How it came about is not clear, but the two of them became involved in a camp for young children at Howth. Half a century later, when the BA was due to meet again in Dublin, he clearly had this earlier visit in mind. The Christmas cards which he sent to his friends in 1956 carried individually attached photographs showing himself—looking almost unbelievably young—and Simonsen sitting in the entrance of a bell-tent with a dozen small boys and girls. Simonsen was interested in Robert's synthesis, and used it himself to prepare terebic acid and two other products, terpenulic and homoterpenylic acids. This work Simonsen published in the *Journal of the Chemical Society* in 1907, but at Robert's request his name was not mentioned. Nevertheless, this simple experiment carried out at home can perhaps be regarded as his first original contribution to organic chemical synthesis.

First-year chemistry at Manchester was then, as was to be expected, of a very general nature, and the turning point in Robert's studies came in the second year, when he first attended the lectures of W. H. Perkin Jr. The simplicity and elegance of these impressed Robert enormously, and he came to regard the system of organic chemistry—at once highly complex and strictly logical—as one of the greatest achievements of the human mind. During this year there was also a course of lectures on the history of chemistry by Bone, and this Robert always regarded as an important part of an undergraduate course—provided it stuck to chemistry and avoided sociology, which he abhorred. Most students also took a course in German, an acknowledgment of the pre-eminence of Germany in the field of chemistry at that time. The final year included advanced lectures on organic chemistry and a choice of special lectures. Robinson elected for a course of lectures, and complementary practical work, by J. F. Thorpe. The latter was another colourful member of the department, to which he had been appointed in 1895, after two years in Germany during which he had worked in the laboratory of a dyestuffs manufacturer. He was a dexterous and versatile experimentalist, whose technique Robinson much admired. He was also possessed of a keen business sense, and gained some profitable consultancies with local dyestuff manufacturers.

The chemistry courses of those days attached much importance to practical work in the laboratory, and students were required to spend long hours there. Some of the work was synthetic—preparing pure samples of specific substances—and some analytical, involving identification of unknown specimens. This was known as 'Spotting', and, in the absence of the many aids which are today taken as a matter of course, involved much intuition and ingenuity, as well as formal knowledge. Laboratory work in the final year was entrusted to D. T. Jones, who invited Robert to collaborate with him in preparing derivatives of bicyclobutane—substances containing two rings, each containing four carbon atoms. Until comparatively recently it had been supposed that only five- or six-numbered carbon rings could exist, four-numbered ones involving too much internal strain for stability. However, Perkin had disproved this by preparing cyclobutane derivatives in the 1880s while working in Germany. The instability argument applied with much greater force to bicyclic rings, and Robinson was not unduly disappointed or surprised when their attempt was unsuccessful. In fact, bicyclobutane derivatives were not made by anybody until many years later.

At the end of the third year there was a wide-ranging written examination covering the whole of the course, and a practical examination extending over three full days. In the latter Robert nearly came to grief, for in writing up his results he forgot that he had left a flame burning under the main product the students were meant to prepare. Whether or not this was a rare lapse one cannot say, but a presumably unbiased laboratory record of 1902 records 'Robinson, Robert—A good worker but messy.' The external examiner was F. S. Kipping, an old friend and colleague of Perkin, with whom he had written a highly regarded textbook *Inorganic and Organic Chemistry*. At that time Kipping was Professor of Chemistry at Nottingham, and was working on a problem requiring considerable quantities of a substance known as menthylamine, and it was this that he had set the students to prepare: this, rather than the allocation of examination marks, may well have prompted his concern about the fate of Robert's unattended specimen.

In the event, the result was all that he could have desired: in 1905, although the youngest person who had ever sat the examination, he was placed head of the list equally with his friend Sandiford. In July his parents came to Manchester to see the B.Sc. degree conferred on him. Looking back, he saw the hand of providence in this: had he been poorly placed through doing badly in his practical he would most probably have given up chemistry and gone into the family business. Instead, he was invited to become a postgraduate research worker in Perkin's private laboratory, an offer which he accepted with alacrity. He was encouraged by a £50 scholarship and the award of the Le Blanc medal.

With this change in his status, and the prospect of an indefinite stay in Manchester, he looked for more suitable accommodation.

The Moravian College was near Buxton, and some twenty miles from Manchester, thus necessitating daily commuting. For a time he shared digs with W. T. Waugh in Manchester; but on graduation their ways parted. In the end a family solution was reached. After a spell at Fulneck his brother Victor, six years his junior, was sent to Manchester Grammar School, and then to the Manchester Technical School for Cotton Spinning and Weaving. The two joined forces and shared an apartment at Buxton.

Before taking leave of Waugh, we must mention an early romantic attachment. Among Robert's fellow students was a botanist, T. G. Bentley Osborn, about his own age, who also stayed on to do postgraduate work, and eventually (1908) was appointed Lecturer in Economic Botany. Osborn became engaged to Kathleen Kershaw, another botanist, whom he married in 1912. While he was at Fairfield Robert had known both Kathleen and her younger sister Ada, of whom he became very fond. Unfortunately, the feeling was not reciprocated, and Ada married his friend William Waugh, who eventually took her off to Canada. However, Robert did not lose touch with the Osborns. Bentley was appointed Professor of Botany at Adelaide at the same time as Robert went to Sydney. They met often in Australia, and in the fulness of time found themselves heads of adjacent laboratories in Oxford.

However, chance quickly assuaged whatever disappointment there may have been over Ada. Very soon after Robert entered Perkin's private laboratory an unexpected foreign visitor was announced. This proved to be Chaim Weizmann, who had studied chemistry at Berlin, Fribourg, and Geneva—an ardent Zionist, who eventually became the first President of the new state of Israel: in the immediate context, however, he was an organic chemist looking for a job. Perkin was greatly impressed, and Weizmann was put in charge of the Schorlemmer Laboratory of Organic Chemistry as Lecturer and Administrator. Initially, he had little opportunity for research, but he shortly acquired a private laboratory when the new Morley Laboratory was opened in 1909. He also became a consultant to Clayton Aniline. This made it possible for him to have a few research workers, among them a young woman named Gertrude Maude Walsh. Her home was at Over in Cheshire, where her father, Thomas Makinson Walsh, was a representative for a coal merchant. He had been a pupil at Manchester Grammar School, and had acquired a good knowledge of the classics—a knowledge which shows through in his memoirs, *The Random Recollections of a Commercial Traveller*. Unfortunately, his business acumen was not as good as his literary talent: in the period 1918–21, at the time of his retirement, he put most of his money into Grand Trunk shares, and lost every penny—some £30,000—in its collapse. He had a large family, of whom Gertrude was the youngest. She was slightly older than Robert, and like him had studied chemistry at Manchester. After graduation she had taught chemistry at Manchester High School for Girls, whose headmistress was a Cambridge mathematician, Sara Burstall, who had gained a national reputation as an educationist.

In due course, Robert and Gertrude became engaged, a development which clearly had the approval of the Weizmanns: both were frequent visitors at their house in Willoughby Road, where many happy evenings were spent. Vera Weizmann was a qualified physician, and, like Robert, an accomplished pianist, with a special liking for Debussy. It was the beginning of a lifelong friendship: many years later Robert became a member of the Board of Governors and a Fellow of the Weizmann Institute of Science at Rehovoth, a research institute of international repute.

W. H. Perkin's discovery of mauveine in 1856 presented great opportunities for the chemical industry; but they were ones which Britain failed to grasp. Although a synthetic dyestuffs industry was established, largely in the Manchester area, by the turn of the century the leadership had largely passed to Germany, whence Britain—like the rest of the world—imported most of the dyes she required for the textile industry. W. H. Perkin Jr felt no urge to make a direct contribution to the advancement of the new industry which his father had founded. Apart from his early research in Germany on cyclobutane, already mentioned, virtually all his work—embodied in more than 200 published papers—was on deriving the structures of natural compounds. These included terpenes and camphor; natural dyes such as brazilin and haematoxyline; and a considerable number of alkaloids, including berberine, strychnine, and brucine.

As a matter of course it was in this field that Robert himself became involved, and Perkin proved an exacting and rigorous mentor. On four mornings a week he lectured to the undergraduate students at 9.30, arriving an hour earlier to supervise the practical demonstration—a teaching aid now sadly out of fashion—organized by A. F. Edwards, the laboratory steward. The rest of the day, until about 4 o'clock, he devoted to his own experiments and to supervising and encouraging his research students. His own research seems to have been conducted something after the style of a Mad Hatter's Teaparty. Beginning at one end of a bench well-stocked with clean apparatus, he worked along it, leaving a trail of dirty apparatus behind him to be tidied by a washing-up boy. By the end of the day, when he left for home, there was usually little space left. Many of his preliminary experiments were conducted on what (by the standards of those days) would be regarded as a modest scale; but once the problem began to resolve itself large quantities—perhaps several kilograms of initial reagents—would be used. This was an approach which Robert himself commonly adopted in much of his later work.

When he took Robert into his laboratory Perkin was particularly interested in two substances—brazilin and haematoxylin—and their complex derivatives. The former is so called because it is the active principle of Brazilwood, one of several species of *Cesalpinia*, much valued as a source of red dye in the days before the advent of the synthetics. Brazilin itself is colourless, but is easily converted by oxidation into the deep red brazilein. The very closely related haematoxylin, from logwood, is derived from several species of trees of the genus *Haematoxylum*. Such vegetable dyes would, of course, have been very familiar and readily available in a great textile centre such as Manchester. During his postgraduate years there Robert published a number of papers on this subject with Perkin, and brazilin had a remarkable fascination for him: it was, indeed, a lifelong fascination, for he published his last paper on brazilin in 1974, only a year before he died. With Perkin he investigated also a number of alkaloids, including strychnine, berberine, harmaline, and harmine. These are complex natural bases found in a variety of plants: many are extremely poisonous. The first of them, strychnine, also held a lifelong interest for him, and at the time of his death he was occupied in writing a monograph on it and some related alkaloids. With Simonsen, he investigated aloes, the dried extract of various species of *Aloe*, long used in medicine as a powerful laxative. The commercial product includes a number of complex organic substances, notably barbaloin.

Although they fell into different classes, the investigation of the wood dyes, the alkaloids, and the aloins all involved structural organic chemistry. The object of the research was to break complex molecules down into recognizable smaller fragments, and from the structures of these deduce those of the original complexes. The final test was to synthesize the original by a series of controlled reactions whose course was indisputable. To generalize, it is probably fair to say that Perkin had no great interest in how these transformations took place: his concern was that the structures of a new substance should be unequivocably related to that from which it was derived. He was a demanding teacher in the practical manipulation of organic compounds, but did not concern himself with the underlying mechanisms. Robinson's interest in this theoretical field—in which in his later years he believed, probably mistakenly, that he made his greatest contribution to chemistry—was stimulated by the arrival in the laboratory of Arthur Lapworth. Lapworth had been a student of H. E. Armstrong, at what was then the Central Technical College, South Kensington (absorbed into Imperial College in 1911, when Armstrong was compulsorily retired). He was a great educationist, and a powerful advocate of the heuristic method. A controversial character, he was a curious mixture of radical and conservative. He was one of the first to use Mendeléev's Periodic Table as a basis for student courses, and perceived that crystal structures in some way reflect the shape of their constituent molecules, a concept later developed in the technique of X-ray cyrstallography by Max von Laue in Germany and the Braggs at Cambridge. On the other hand, he ridiculed Arrhenius's theory of electrolytic dissociation, the fundamental importance of which earned the latter a Nobel Prize in 1903. Thus Lapworth was trained in a stimulating but sometimes stormy atmosphere. In 1900 he moved on to become head of the chemistry department at Goldsmith's College, London, and in 1909 he moved back again to Manchester.

Arthur Lapworth and Robert quickly formed a warm and lasting friendship based on a variety of mutual interests, both personal and professional. Robert already knew Armstrong, having stayed with him for a few days, soon after his appointment at Manchester, at a house he rented on Derwent Water. He formed a considerable liking and respect for him, though he was somewhat contemptuous of his carrying an ice-axe while fell-walking on fine days. It reminded him, he said, of Mark Twain's ascent of the Riffelberg. No doubt Armstrong's vagaries were discussed when Arthur and Robert went on their many walking and climbing excursions together. They also had a common interest in music. Robert's talent as a pianist was matched by Arthur's as a violinist; and they shared a liking for Mozart. Arthur was also interested in birdwatching, and could identify all the local wild birds by their song—an interesting accomplishment, but one which Robert recalled as being rather wasted on him. The identification of wild flowers, however, aroused the enthusiasm of both of them equally. An easy personal relationship fostered the development of an important joint chemical interest. Before he came to Manchester, Lapworth had developed a theory of alternate polarities in organic molecules as an explanation of their behaviour in a variety of reactions. Originally, this was little more than a useful mnemonic; but together they developed it into a sophisticated electronic theory of organic reactions. Their colleagues in the laboratory became accustomed to the sight of 'Lapworth and Robinson playing noughts and crosses together' as together they worked out the consequences of the theory. In later years their advocacy of this theory was to lead Arthur and Robert into fierce public controversy; but this is a tale that must be left to a later chapter.

Lapworth, Perkin, and Kipping had a curious family relationship. In 1887 Perkin married Mina Holland, daughter of a Liverpool physician. Kipping married her sister Lily, and Lapworth a third sister, Kathleen. Thus the three Holland sisters all married organic chemists, who all became Fellows of the Royal Society.

The years at Manchester passed away pleasantly enough. For a year (1910–11) his experience was enlarged, and his finances somewhat improved, by an appointment as Residential Tutor in Chemistry at Dalton Hall, a university hall. Even as such he was not accorded the privilege of a key, which conflicted with his habit of working in the laboratory until the early hours of the morning. This small problem was solved by his mountaineering skill, which made it easy to climb in when he wished. His research was to his liking and going well: thirty-five good papers published in his first seven years, earning him his D.Sc. degree in 1910. His colleagues were agreeable, and many had become his friends, sharing his own interests. At a personal level there was his courtship and engagement—to an attractive young woman who shared his interest not only in chemistry but in walking and climbing; and, despite the sadness of his father's death in 1911, still the background of a happy family home at Chesterfield. There was thus little inducement to make any change. Like many chemists of his day he might well have gone to one of the great German laboratories for a year or two to gain additional experience and perfect his German. That he did not do so was perhaps due, at least in part, to his attachment to Gertrude. His meagre academic salary made marriage difficult; and in those days she could hardly have accompanied him to Germany except as his wife. The alternative of a long separation can scarcely have appealed to either of them. Whatever the reason, he remained at Manchester for a total of ten years, and when he finally made a move it was, at first sight a little surprisingly, to the opposite side of the world.

3
The first Chair

The school of organic chemistry presided over by W. H. Perkin Jr at Manchester from 1892 until he was appointed Waynflete Professor at Oxford in 1912 was internationally famous, even in comparison with the great German schools, such as that of Emil Fischer in Berlin. To have been appointed Assistant Lecturer and Demonstrator there at the early age of twenty-three augured well for Robinson's future: he could reasonably expect to climb steadily up the academic ladder in Britain, and in the fullness of time achieve one of the more prestigious professorships—which, indeed, is what he did. At first sight, therefore, it is something of a surprise to find him applying, in 1912, for a newly created chair in the University of Sydney, literally on the other side of the world. Whatever the attractions of Australia it was undeniably remote, and cut off from the main stream of academic activity: the journey out from Britain occupied several weeks, and postal communication took just as long. Even within Australia communications were not good.

To find a plausible explanation we must recognize that conditions seventy-five years ago were very different from what they are now. Britain was still a great imperial power, reluctant to grant more autonomy than was strictly necessary to the constituents of her farflung empire. Only about the middle of the nineteenth century, some sixty years before the time with which we are now concerned, had overseas universities been viewed with favour. Until then, colonials wanting a university education had to seek it abroad: in practice, a combination of circumstances normally brought them to Britain. The first Indian Presidency Universities (Calcutta, Bombay, Madras) were not set up until 1857; and the first in Australia (Sydney, Melbourne) only a little earlier, in 1850 and 1853 respectively. Of necessity, these new institutions had to look to the old-established British universities—notably Oxford, Cambridge, Glasgow and Edinburgh—for their academic staff. While these home universities were attractive in terms of prestige and quality of life, they were relatively small and static, and offered correspondingly limited opportunities for promotion. Graduates bent on an academic career were, therefore, not too reluctant to take appointments in the new overseas universities, especially as experience showed that this was by no means a one-way traffic: many of them later returned to take up good appointments in Britain. This system of appointing from Britain persisted until after the turn of the century, even though it was then no longer necessary. By that time the new universities were becoming restive, maintaining—though with little reason—that in the course of half a century they had reached a stage of development at which their own graduates were perfectly capable of occupying senior positions. In 1978 P. H. Partridge, Professor of Social Philosophy at the Australian National University 1951–75, wrote: '. . . many people have no conception of what very narrow little parochial oligarchies our universities in many respects were until after World War II.'[1] As we shall see, the Robinsons became aware of this attitude when in due course they arrived in Sydney. The immediate point, however, is that in pre-war Britain it was by no means unusual for a young academic of Robinson's calibre to apply for such an overseas appointment. Indeed, his close friend and fellow student John Simonsen had left Manchester in 1910 to become Professor of Chemistry at the Presidency College, Madras: he was to return twenty years later as Professor of Chemistry at University College, Bangor, and to be elected FRS in 1932. It appears too, that Perkin himself recommended that Robinson should put in for the job.

Apart from those general considerations, there were other factors to consider, not least the fact that the salary—£900 a year plus £100 travel expenses—was far greater than he was earning at Manchester, and more than he could expect to earn at that time in any post at home. But by this own account the main reason lay in none of these considerations but in his lifelong passion for mountaineering:

> I applied for the post in Sydney, in the first instance because it would be an excellent base for the exploration of the alpine chain of the New Zealand Alps. This was my primary objective, not, I am afraid at that stage, the advancement of the science of chemistry.[2]

However, there may have been a little more to it than that, for Australia was not wholly an unknown land in the Robinson household. Robert's father had been there on an extended business tour in 1893/4, and he had distant cousins there at that time. There is, however, no evidence that he contacted them during his appointment at Sydney, though his brother Victor did so later during visits in the 1930s and 1950s.

In 1827 Maria Robinson, married Benjamin Boothby, whose father had a large iron-works at Furnace Hill, Boythorpe. Benjamin had been called to the Bar in 1825, and at that time practised on the Northern Circuit. In 1853 he was appointed Judge of the Supreme Court of South Australia, and with his wife and twelve children went to Adelaide. He died in 1868, but his wife survived him until 1889, when she died in her eighty-fourth year: she was the last survivor of William Robinson's large family. The further ramifications of the Australian Boothbys are of no direct interest in the present context: they included engineers, civil servants, farmers, electricians, and accountants.

The Sydney post was advertised in March 1912, and in June Robinson dispatched his formal application: in the liberal tradition of those days it was printed, not typed. In support he listed no less than 30 papers already published in the *Journal of the Chemical Society*, and indicated that more were imminent. Among the authors of the appended testimonials were H. B. Dixon, Professor of Chemistry at Manchester; F. S. Kipping, of University College Nottingham, who was to make a name for himself in advancing the organic chemistry of the intractable element silicon; and Jocelyn Thorpe, Sorby Research Fellow at Sheffield. All three were eminent in their fields, and Fellows of the Royal Society; their support was thus powerful, and they gave it in full measure.

In some respects, the testimony of Dixon might well have been written at a later date, when Robinson had had longer to show his mettle:

> [He] is remarkable among the chemists I know for having both an extraordinary knowledge of the literature of his subject, old and new, and the power of rapid and accurate work along original lines. It is, I think, very rare to meet a man with such power of observation and quick induction able to check and weigh his conclusions with such knowledge and judgement.

In other respects, however, events showed that Dixon's assessment was less sound. He goes on to say:

> I can also bear testimony to the clearness and interest of Dr. Robinson's lectures. He at once arouses an alertness in his class which even the backward soon catch.

With this, few of his former students would agree. For the better ones his lectures were fascinating, but not because they corresponded in any comprehensive and systematic way to the advertised course. His quick mind would lead him away to discuss some complex problem of structure or synthesis that was then preoccupying him. The less bright students were quite unable to follow him. A. J. Birch—who knew him at Oxford for some ten years from 1938, when he arrived to take up an 1851 Research Scholarship, and in due course (1952) came to occupy the same chair at Sydney—has made the point very succinctly:

> He was not, even in 1938, a good undergraduate teacher in the standard sense. He never gave the impression of having prepared a lecture, he just talked on the subject. This was fascinating for the best students, particularly as he was likely to have some new idea suddenly, and start to work it out on the spot. . . . The students had to teach themselves largely by observing how a great mind coped, rather impromptu, with a topic, and this is probably the best way to learn. His formal scientific lectures could vary from being very good and exciting, if he was really interested, to embarrassingly bad if he was not . . . his lecture to the Royal Society of Arts in 1948 [on 'The Structural Relations of Some Plant Products'] is an example of an extreme case where the ordinary members of the Society clearly had no idea what he was talking about.[3]

As his contribution, Kipping remarked that 'it is, in fact, very unlikely that there will be any other candidate having so brilliant a record.' This was evidently the opinion of the assessors, for Robinson was duly appointed.

Robert and Gertrude were married in the parish church at Over, Cheshire, on 7 August 1912, with her parents as witnesses. He was then 25 years of age, and she 26. They set out for Australia in early December, and they made a leisurely progress, passing a week in Chamonix before spending Christmas in Florence. This was not their first visit to the Alps that year, for they had done some strenuous climbs in the summer, including Mount Collon and the Aiguille de la Za. From Florence they went on to spend a week in Naples, before embarking on a German ship, the *S.S. Scharnhorst*, for Sydney. Again it was a leisurely progress, with stops at Colombo—giving them their first taste of tropical heat, and an opportunity to visit Kandy—and then on by way of Fremantle and Adelaide, where the call of Mount Lofty could not be ignored. In parenthesis, it may be remarked that in Adelaide he must have been 'within coo-ee' of a young Australian with whom he was to be closely connected in later life. This was Howard Florey, pioneer of penicillin, then a schoolboy of 14 at St Peter's Collegiate School, which some of Robert's Boothby relatives had attended. Both were to be Nobel Laureates, Presidents of the Royal Society, and holders of the rare distinction of the Order of Merit. Four years earlier, and the Robinsons might have met W. H. Bragg, professor of mathematics and physics at Adelaide, whose career would illustrate the point made earlier. A Cambridge graduate, he had been appointed Professor of Mathematics and Physics at Adelaide in 1886, at the age of twenty-four. He returned to England in 1908 as Professor of Physics at Leeds, and embarked on a brilliant new career particularly identified with X-ray crystallography, which from 1923 onwards saw him installed as Director of the Royal Institution.

At Melbourne a temperature of 105° gave them a foretaste of things to come. Finally they reached Sydney on 7 February 1913. By the terms of his appointment, he was expected to take up his duties on 1 March, but the University hoped that the new professor would arrive in Sydney some little time before the start of the academic year. This no doubt suited the Robinsons well: it gave them a little leisure to find somewhere to live and, no less important, the professorial salary commenced on the day they arrived.

Before continuing, it is appropriate to digress for a moment and consider the kind of community they were joining. Sydney, founded in 1788 as a British settlement for the convicts originally destined for Botany Bay, was blessed with a magnificent harbour; a temperate sunny climate; and sufficient rainfall to ensure an adequate water supply. Thus favoured, it prospered, and in 1849 its population was around 45 000, a quarter of the population of New South Wales. At that time its only institute of higher education was Sydney College, a private high school, but this was on the point of closure. Young men wanting a university education had to seek it in Britain, 12 000 miles away. Parents feared, and no doubt with reason, that at such a distance the benefits of home life might be forfeited without 'any compensating improvement to their minds'. A Bill to establish a university was passed in 1850, and the first students—no more than 24—were enrolled in 1852. Its subsequent history need not concern us in a general way in the present context, but the history of the chemistry department is very relevant to the situation in which Robinson found himself. Chemistry and Experimental Philosophy had been one of the foundation Chairs, but over the years the department had become oriented to the needs of the mining industry: the New South Wales coalfield was the largest in the southern hemisphere, and from the 1880s Broken Hill had been famous for its deposits of silver, lead, and zinc. In 1909 the Senate appointed to the Chair Charles Fawsitt, then Lecturer in Metallurgical Chemistry at Glasgow. He was well qualified, having studied not only at Glasgow but under Wilhelm Ostwald at Leipzig, but he was essentially what we would now call a physical chemist. Although organic chemistry had been taught since the 1880s, the facilities were limited. It was against this background that Fawsitt recommended, soon after his arrival, the establishment of a Chair of Organic Chemistry, but the State Government would not authorize a second chair of pure chemistry. As a compromise, a Chair of Organic Chemistry—Pure and Applied, was agreed upon by the Senate. It was doubtless because the Chair was so designated that Dixon, in his testimonial, stated:

'Dr Robinson has a good knowledge of many applications of Chemistry, and has made a special study of the theory and practice of Dyeing, on which he is our chief lecturer.'

Though made in a good cause for a young colleague, this statement was somewhat stretching the truth, as was a statement in the 1913 Sydney University Calendar that he was 'The author of many original papers on chemical subjects, including contributions to the chemistry of dyeing processes.' Although he did indeed have a lifelong interest in the chemistry of natural dyes and pigments, his knowledge and experience of dyeing as an industrial art was minimal. A cursory glance at the titles of the published papers submitted with his application shows that none had any reference to the practical use of dyes. But no matter: Sydney had acquired a professor of organic chemistry—the first in the southern hemisphere—of great distinction. Their misfortune was that his stay was so short; but this, in the context of the time, was inevitable. In more recent years, when facilities and communications have been transformed, it might have been a different story.

Once arrived in Sydney, the Robinsons' first need was to find somewhere to live, and here they had immediate good fortune. Little more than a year previously Wyoming Professional Chambers had been completed, on the corner of Macquarie and Hunter Streets. An eight-storey building nearly forty metres high, it was one of the landmarks of the city. As its name implies, it was designed to provide accommodation for professional people, in the event mainly doctors and dentists. There were over a hundred rooms, but the top floor alone provided residential accommodation: one flat, with a roof garden, and accommodation for the caretaker, Samuel Lancaster. Robinson recalls that, when showing them round, the latter strongly recommended the flat, because of the splendid view it afforded of the constantly occurring fires in the Parramatta. In a different direction it looked out over the Botanic Gardens and the harbour. It was some three miles from the University, but the splendid position must have more than compensated for this: for such dedicated walkers this was no more than a stroll. The Robinsons remained there throughout their time in Sydney.

A few days after his arrival Robinson was interviewed by a reporter from the Sydney Morning Herald, in the course of which he is reported as saying that he:

> . . . was much impressed by the fine [University] buildings and their beautiful situation. They were much finer than anything he had before seen. The chemical laboratories . . . were very well appointed, and he had never seen a school or University library to equal that of the Sydney University.[4]

If accurately reported, he must have exercised some tact. His own personal accommodation was in fact very limited—a research laboratory measuring 7×5 metres in the basement, with some additional space for research students. At the time, this need not have been too disappointing, for there was provision for a new building to house organic chemistry. Although he designed this new laboratory, he never had the advantage of it, for the plan was shelved on the outbreak of war in 1914. The first beneficiary was his successor, John Read, from Cambridge, who was later to succeed Robinson a second time, in the professorship at St Andrews.

Apart from archival material at Sydney, there is by chance a generally available description of the chemical laboratories as Robinson knew them. In his *Chemical Discovery and Invention in the Twentieth Century* William Tilden singled them out 'to show what has been done in a distant part of the British Empire.'[5] The main laboratory, providing for all students except beginners (mainly medical) was a lofty room some 24×12 metres: the assay laboratory, largely for metallurgical work, was roughly twice as big. The large lecture room could accommodate 200 students—about the normal total at that time, including medical and engineering students doing part-time courses—and there were separate rooms for balances, spectroscopes, and polariscopes. Clearly the facilities there were much as he might have expected in Britain. In addition to Tilden's book, which concerns itself with chemistry, two other works throw much light on the University of Sydney as it was in Robinson's day. Both are commemorative works, one a general history of the University 1850–1975,[6] and the other marking the centenary of the awarding of the first degrees in science.[7]

The new post carried a fairly heavy teaching load. While his new students in the science faculty apparently presented no problem, he had also to do service teaching for medical and engineering students, whose interest in organic chemistry was minimal. In their view these absurdly detailed courses were irrelevant:

What does a third year Med. remember of his first year subjects? Zoology? Nothing, save a little vertebrate anatomy. Botany? Nothing whatever. Physics? One or two general principles, which have had a later physiological application. Chemistry? The same. All the wearily swotted details that made first year a nightmare are gone, and he is glad to be rid of them. Yet upon the teaching of those details all the energies of the senior staff are concentrated.[8]

At Adelaide, at much the same time, the young Florey, too, was resentful of this rigid approach. In later life he recalled students being told: 'It is far better for you to learn what is in your books. It is not for you to question.'

With this attitude it is not surprising that the medical students tended to be noisy and unruly. Robinson's predecessors had allowed them to get out of hand: missiles were commonly hurled about, and on one occasion, at least, a hurdy-gurdy was brought into a lecture. This kind of behaviour was, of course, quite outside Robinson's insular experience; but he seems to have coped fairly philosophically: however, his recollection that 'he usually managed to get a hearing' suggests that there were times when he did not. Gertrude seems to have got the students' measure very quickly. Their laboratory had two doors, one opening on to a corridor and the other on to a chemical collection room, and it had become the habit for students to come in through one and out through the other in pursuit of some kind of gang warfare. She put a stop to this by simply saying 'You cannot come here, this is Professor Robinson's private laboratory'.

As we have noted, by the turn of the century there was growing discontent about the traditional practice of making senior appointments from the UK; it was felt that many of these could by then be given to Australians who had graduated at their own universities. This view had been well ventilated in the student magazine *Hermes*:

> What we do protest against is the deliberate rejection of our own men, though thoroughly suitable, in favour of others who are comparatively unknown. It is, we repeat, a discouragement, to scholarship in general, to University men in particular, and a belittling of the work our Universities are doing . . . It is our deliberate opinion that the great disadvantage with which scholarship here has to contend, is not the great distance from other centres of culture, not the lack of great libraries, but the fact that it is next to impossible to secure recognition from those who might expect to foster learning and culture.[9]

However, this discontent seems not to have been at all directed at Robinson personally, who with typical Australian lack of formality was amicably referred to as 'Prof. Bob Robinson'. Like other students, they had difficulty in keeping up with his two new courses of lectures, modelled on those at Manchester:

> He deals out organic chemistry at race-horse speed. He wants badly to tell you all he knows; but it is hard to do it in an hour when one knows so much. He has been learning chemistry all his life.

They evidently had a good-natured attitude also towards Gertrude:

> He has a wife to whom one may apply for information, if he himself is not available. He told us so. Yes, by all means refer to Mrs Robinson.

This was fortunate, for she occupied a somewhat anomalous position in the department. From its foundation the University did not specifically exclude women, but in fact none enrolled for matriculation before the 1880s, and in the Robinsons' time—and long afterwards—women gave up their academic jobs when they married. Even Jane Russell, Tutor to women students, had had to resign when she married Henry Barff, the Registrar, in 1899. Nevertheless Gertrude, already married when she arrived, was made a Demonstrator in Organic Chemistry. However, this preferential treatment does not seem to have prevented a warm friendship being established with the Barffs.

During his time at Sydney Robinson prepared about twenty-five research papers, several jointly with Gertrude. Some were write-ups of work done at Manchester before he left, and they were mostly on natural products, notably alkaloids. In this field he developed a new interest through contact with an unusual man, Henry Smith. In England he had been a sign-writer by trade, and had come to Australia for reason of health. Through evening classes, he had gained a good knowledge of chemistry, and had been appointed Museum Mineralogist. His real chemical interest, however, was in the chemistry of the essential oils of Australian plants, especially eucalyptus oil, on which he became a leading authority. This had a natural appeal for Robinson, and together they published several papers, including one on eudesmin, an interesting eucalypt constituent.

Just as it had been decided to strengthen the chemistry department by creating a new chair of organic chemistry, so biology had at the same time been strengthened by a new chair of botany. The man appointed, Anstruther Abercrombie Lawson, arrived in Sydney about the same time as the Robinsons. He was a Canadian of Scottish parentage, and, like Fawsitt, was appointed from Glasgow University. Apart from both being new arrivals together, he and the Robinsons found kindred interests—professionally in the classification of plants, and socially in walks around the Blue Mountains and elsewhere. Lawson had one other interest which intrigued Robinson. He believed that many of the early Scottish immigrants to Sydney brought among their domestic luggage some valuable Old Master paintings, and he was constantly on the lookout for unidentified treasures of this kind, aspiring even to a Rembrandt. Whether his hopes were ever realized we do not know.

Organic chemical experiments carry a particular risk of fire, because of the frequent use of highly flammable solvents such as ether and benzene. During their stay in Sydney the Robinsons suffered a serious setback through a fire which gutted their laboratory and destroyed all the chemical specimens which they had accumulated up to that time. This was a great loss, for they were the basis for comparison with new products, and in the nature of things replacements were not commercially available. Of less serious consequence were fires that started during student experiments, but Gertrude gained a reputation as an efficient firewoman:

> Mrs Professor Robinson, who acts as an honorary demonstrator for her husband, and says the things to the students that he is not game to say, is an amazing amateur fire brigade, all by herself. Fires are of frequent occurrence in organic chemistry, and a rug is specially provided with which to

extinguish them. When a blaze arises Mrs Robby hurls herself into the air, grabbing the rug as she flies, falls upon the conflagration, puts it out regardless of singed hair and eyelashes, and, without even waiting to regain her breath, rounds upon the luckless student until he wishes the fire had consumed him too!

The year 1914 was a doubly significant one. At the national academic level it was enlivened by the Annual Meeting of the British Association for the Advancement of Science. This body, less influential now than it was then, organized week-long annual meetings at which scientists could describe recent progress in their own field not only to their peers but to members of the general public. They were large gatherings, often numbered in thousands, and were very widely reported in the press. Normally the meetings were held in university cities in Britain, but very occasionally they were held in the Commonwealth. The 1914 meeting was the first truly international scientific conference to be held in Australia, and was a protracted event, beginning at Perth before proceeding eastward to end at Sydney. The UK contingent included such distinguished chemists as W. J. Pope, J. E. Thorpe, and N. V. Sidgwick.

For the Robinsons this was an event of major importance and, as we shall see, the impression Robert made on some of the distinguished visitors was a material factor in his subsequent appointment to a new chair at Liverpool. At the time, however, it must have been a very fraught occasion. According to the programme they were each to have given a paper to the Chemistry Section on 21 August: Robert on *Researches on the Synthesis of Isoquinoline Alkaloids* and Gertrude on *The Condensation of Cotarnine and Hydrastine with Aromatic Aldehydes*. According to the *Sydney Morning Herald* of 22 August they did in fact give these lectures. They were, however, described as being 'of a highly technical character', and we must suppose that their reporter was daunted and relied on the printed programme and did not attend in person. Sadly, Gertrude, who had become pregnant early in the year, lost the baby on 20 August; a daughter, she lived only one day. It seems that Robert delivered his own lecture, and possibly also read Gertrude's, but just how matters were arranged to meet the crisis is not now clear.

This personal tragedy was overshadowed by a far greater international one, for during the course of the meeting war broke out in Europe. Remotely situated as it was on the other side of the world, the immediate consequences for Australia were not great; but in the long run they were appalling. The still young country, with a population of less than six million, sent 350 000 troops overseas, of whom 60 000 never returned: all this without introducing conscription. By 1916 the University had shrunk to little over 400 students, of whom 129 were women: in all some 1700 members of the University enlisted, including nearly all the graduating class in medicine and engineering. Robinson found himself involved in the activities of the University OTC, and underwent some military training at a camp at Kiama, on the coast a hundred miles south of Sydney: he proved himself a tolerable marksman.

Despite this disruption opportunity was found for mountaineering trips to New Zealand. Successes there included the scaling of Coronet Peak—sometimes used as a training ground for aspiring Everest climbers—under the guidance of Conrad Kain, chief guide of the Canadian Alpine Club; the first ascent of Mount Meeson; and a double traverse of Mount Egmont, recorded as having 'astonished the natives quite considerably'.

The British association visit was a welcome stimulus, and a means of renewing contact with leading chemists in Britain. At a more personal level he kept in touch with family affairs through correspondence with his sister, Dorothy, then eighteen years old and still living at home. One of his few surviving personal letters, dated 15 December 1913, is addressed to My dearest Dorothy', and thanks her for her many letters to the 'exiles' in Australia: characteristically, he adds that he is 'just off to New Zealand on a climbing trip'. Whether at that time he made contact with any of the Boothbys is not known, but bearing in mind that Benjamin Boothby had emigrated 60 years earlier they were by then no more than distant relatives. However, his brother Victor established a close friendship with a later Benjamin Boothby in the 1930s.

Thus on the whole life passed agreeably enough. The research went well; they were comfortably housed in a city they liked; they made a circle of friends; and they had good opportunities to indulge in the walking and climbing that meant so much to them both.

Then, quite suddenly, came a new development which was to change the whole pattern of their lives. In 1915 the University of Liverpool created a new chair of organic chemistry, endowed by Sir Heath Harrison, a wealthy local shipowner. The post was duly advertised, and Robinson applied. It attracted ten applicants, including such strong contestants as T. S. Patterson, who later became professor of organic chemistry at Glasgow, and (Sir) Robert Pickard, who subsequently had a distinguished career as Director of the British Cotton Industry Research Association and Vice-chancellor of London University. Nevertheless, only two other candidates were short-listed with Robinson. They were K. J. P. Orton, who already held a professorship at the University College of North Wales, where he was elected FRS in 1921, and Samuel Smiles, then assistant professor at University College London, and subsequently professor at King's College London, elected FRS in 1918. Both were considerably older men—Orton was forty-three and Smiles thirty-eight—and their occupation of chairs elsewhere, and the fact that they were in due course elected to the Royal Society, is a fair indication of their calibre. Against this Robinson was a mere twenty-nine, and, moreover, was not available for interview.

Nevertheless, the powerful support of his distinguished referees carried the day.[10] They were led by W. H. Perkin of Oxford, who wrote:

Robert Robinson has, since 1905, published a long series of investigations whch are characterised not only by extraordinary originality and theoretical insight, but also by experimental ability of the highest order, and it is remarkable that so young a chemist should have found it possible to have carried out so much work of this magnitude in so short a time . . . An excellent lecturer, very clear and concise, he possesses the power of preserving [sic] his subject in an attractive and often quite original light which always commands attention.

The importance of the British Association's visit to Sydney in the appointment is apparent from the testimony of N. V. Sidgwick, another eminent Oxford chemist and a man not given to undue praise:

My personal acquaintance with him is derived only from the week that I was with him at the British Association in Sydney last August: but I saw enough of him then to be very much struck not only with his great scientific capacity and energy but also with his very remarkable personal charm—no other word will describe it . . . I have seldom if ever met a man who produced so great an impression on me so soon.

Another visitor to Sydney at that time was H. E. Armstrong, Professor of Chemistry at the City and Guilds College, and well known for his advocacy of reform in scientific education. He was not a formal referee, but nevertheless chose to write to the Vice-Chancellor at Liverpool:

I not only know what his career has been and his published work, but saw him recently in Sydney . . . I am clear there is no one else here who in any way approaches him in ability as an organic chemist.

After due deliberation the Committee concluded that 'the claims of Professor Robinson were decidedly superior to those of the other two candidates, and unanimously agreed to recommend him for the appointment'. This was approved by University Council on 13 July 1915, and Robinson was informed accordingly: the letter of appointment confirmed that his stipend would be £600 a year, less than he was receiving in Sydney.

His resignation from Sydney was effective from 31 August, which left the Robinsons ample time for a leisurely journey home to take up his new post in October. Because of the war, they sailed east—via Auckland, Suva, and Honolulu—to Vancouver. Their ship was the 7500 ton RMS *Niagara*, which left Sydney on 2 September with 250 passengers on board: the Robinsons travelled in comfort in the First Saloon.[11]

The attraction of the Canadian Rockies was not to be resisted, and they spent some time at Lake Louise before going on to New York via Buffalo and Niagara Falls. Arriving finally at New York they were reminded of the reality of war: the liner which was to have taken them on to Liverpool had been sunk by a German submarine. After a delay of ten days they embarked on the *St Paul*, an old ship of the US Navy—festooned with lights to proclaim its neutrality—in which they finally reached Britain without incident.

For the moment interest centres on Liverpool and their new life there. It is opportune to remark here, however, that 1960 found Robert back in Sydney for a symposium on the chemistry of natural products organized by the International Union of Pure and Applied Chemistry. On that occasion his old university took the opportunity of conferring on him on honorary D.Sc. degree.

4

Liverpool and a flirtation with industry

In a sense, in going to Liverpool Robinson was returning to his alma mater, for when he first went to Owens College in 1902 this was associated with University College, Liverpool, and Yorkshire College, Leeds, as the federal Victoria University of Manchester. In his second year a reorganization took place, and Leeds, Liverpool, and Manchester were granted individual university charters. He was also returning to familiar ground, for Liverpool was only some sixty miles from his native Chesterfield, and only half that distance from Manchester: Gertrude was even nearer her old home at Over, near Northwich.

If the war had seemed rather remote in Australia, it was certainly not so in Britain, where hopes of an early peace had been abandoned. The Ministry of Munitions had been established to marshal all industrial resources for the support of the armed forces: by the end of the war it controlled three million workers. The Department of Scientific and Industrial Research (DSIR) was set up similarly to mobilize scientific skills. Bloody and inconclusive battles had been fought in France; raiding Zeppelins had appeared over London; at sea nearly half a million tonnes of shipping had been lost in the first half of 1915, including the *Lusitania*, sunk in May with the loss of 1200 lives. What seemed a promising new initiative in the Dardanelles, in which the Australian contingent played a distinguished role, turned out badly, and was abandoned at the end of the year. Some of Robinson's earlier colleagues from Sydney served there, as did his cousin Theo, who was in the RNVR.

Within his own personal circle, the war had brought many changes. His brother Victor had been gazetted Second Lieutenant in the Sherwood Foresters, and in November 1915 won the Military Cross for bravery during the attack on the Hohenzollern Redoubt. He was later twice wounded, but survived to bring the cadre of his battalion back to Chesterfield for demobilization, by which time he had risen to be Lieutenant-Colonel.* These must have been anxious days for Robert, for he and Victor had always been particularly close. Florence and Dorothy were both at the Endell Street Military Hospital in London, Florence as a nursing orderly and Dorothy in the X-ray department. Cecily, too, was away, serving as a V.A.D. in a French hospital in the Rheims sector.

His half-brother William was twenty-three years Robert's senior. As a young man he had been a keen member of the Volunteers, and had served with distinction in the South African War. Although he continued to support the newly formed Territorial Army, ill health obliged him to retire from it in 1911. In 1916 he succeeded his uncle, Charles Portland, as Chairman of Robinson and Sons.

There had been changes, too, at Manchester University. W. H. Perkin Jr, who had encouraged and supported Robinson in his application for the Liverpool chair, had in 1912 moved on to the Waynflete Professorship of Chemistry at Oxford, where Robinson was destined to succeed him in 1930. Perkin's place as Professor of Organic Chemistry had been taken by Lapworth, with whom Robinson was able to resume discussion of their electronic theory of organic chemical reactions.

Their old friends the Weizmanns were still based in Manchester, but Chaim was about to leave to become director of the Admiralty Laboratories. Behind this somewhat surprising appointment, seeing that Weizmann was a biochemist, lies an interesting story, which had far-reaching implications. One of the essential constituents of high explosives was acetone, but existing methods of production were quite inadequate to meet the need for it as the demand for shells, bombs, torpedoes, and other military weapons soared. The Germans were known to be making it from acetylene, but their patents proved very uninformative, and in practice unworkable. In the course of research on the manufacture of synthetic rubber Weizmann had isolated a bacterium (*Clostridium acetobutylicum*) which could ferment grain to produce a 1:2 mixture of acetone and butyl alcohol. This was brought to the notice of the Admiralty in 1915, and from early 1916 increasing amounts of acetone were produced by this new process. This important development attracted the interest and friendship of A. J. Balfour, who was no stranger to industrial chemistry: in the 1880s he had, for example, been closely concerned with the formation of The Aluminium Company to exploit H. Y. Castner's new process for making cheap sodium. In 1904 he had been President of the British Association for the Advancement of

Science. In 1915 he was made First Lord of the Admiralty, and remained there until the end of 1916, when he became Foreign Minister under Lloyd George. Balfour was thus a very influential man, and he was anxious to acknowledge Weizmann's extremely important contribution to the war effort. The latter chose as his reward not a personal distinction but the famous Balfour Declaration of 1917, promising British support for a Jewish national home in Palestine. This had far-reaching consequences. Weizmann was for many years President of the Zionist Organization and the Jewish Agency. When his life's dream was realized in 1948 with the creation of the state of Israel, he was its first President. With these essentially political developments in his life Robinson had no direct concern, but, as we shall see later, he and Weizmann had a strong and lasting association in the field of science. Weizmann was the first director of the Sieff Research Institute, founded in Palestine in 1934, which subsequently became the internationally famous Weizmann Institute of Science, Rehovot: of this Robinson became an honorary Fellow, and he visited it often.

Robinson took up his new appointment in January 1916, and he immediately resumed his research on the structure and synthesis of alkaloids—two papers in 1916 and eight in 1917—and gave a regular course of student lectures. We now have no direct recollections of him as a teacher at Sydney, but David Nealy, who was a chemistry student at Liverpool 1917–20, still remembered him well in 1985:

> He had a most attractive personality with a very modest element which often showed itself. His lectures were clear and fascinating and memorable. I can still recall his lecture on the elements of organic chemistry and that he told us of his own researches and interests, which at that time were the synthesis of the alkaloids.

He goes on to recollect the personal interest he took in his students:

> On one occasion he left his class some written work—principally how to convert A to B via Grignard reaction+allylacetaldehyde etc. When he returned our rather moderate efforts, we found they were carefully corrected in red ink, with copious annotations. . . . I remember his gifted wife Gertrude, who worked with him and the happy relations with his colleagues, especially Prof. E. C. C. Baly.[1]

It says much for Robinson that he could maintain this interest in the face of unusual demands. The technical problems of wartime Britain could not be ignored, and, as in other universities around the country, Baly, Professor of General Chemistry and in effect head of the department, had organized students and staff to manufacture various chemicals in short supply. Quantities of the local anaesthetics novocaine and beta-eucaine were made for the National Health Insurance Committee, and Robinson initiated an investigation of the possibility of synthesizing the mydriatic drug atropine. Of more direct military interest was research on the high explosive picric acid and some of its by-products. He and his colleagues also investigated processes used in the manufacture of TNT (trinitrotoluene), and Robinson records in his *Memoirs* how bucketfuls of highly dangerous sludge from the Silvertown Factory—severely damaged in a catastrophic explosion in January 1917—were kept under the laboratory benches.

As we have remarked, Robinson had a lifelong interest in the chemistry of natural products, but at Manchester and Sydney he had been concerned largely with establishing their chemical structures by degradation and synthesis. While at Liverpool, however, he increasingly turned his attention to the mechanism by which such substances are formed in nature. This was necessarily somewhat speculative, and he quickly decided that it was more appropriate to refer non-committally to structural relations, rather than use his initial term biogenesis, with its more specific implications.

Unlike many academics of his generation, who cut themselves off from the world in their ivory towers, Robinson enjoyed the busy and stimulating world of industry and commerce, and made many friends within it. With his industrial family background, this need be a matter of no surprise. One such contact takes us back into the folk history of the British chemical industry. Its acknowledged founder was James Muspratt, born in 1793, who set up an alkali works on Merseyside in 1823. In 1837 he began a long friendship with the great German chemist Justus von Liebig, who had come to Liverpool for the annual meeting of the British Association for the Advancement of Science. As a result, Muspratt sent three of his sons to study under Liebig at Giessen, and himself often visited there. Of these, Edmund Knowles Muspratt, born in 1833, was still living in Liverpool, where he died in 1923. The Robinsons visited his home, and greatly admired a splendid oil painting of James Muspratt which hung in the study. Edmund was a director (later Vice-President) of the United Alkali Co., which in 1926 became one of the constituent companies of ICI, a company with which Robinson was to have a close association for some thirty years, until he changed his allegiance to Shell.

The University gave active encouragement to collaboration between its own staff and local manufacturers. In the case of chemistry, this was effected through an Advisory Board, set up in July 1917, which included very senior representatives of the local chemical industry—Lever Bros, Castner-Kellner Alkali, Salt Union, Brunner Mond, United Alkali, and Joseph Crosfield. Robinson and Baly, with five of their chemical colleagues, were the other members. Such collaboration could be particularly effective in the Liverpool area, where some two-thirds of the total manpower of the British chemical industry was employed. The Board's Report of February 1918 stressed that:

> it is now our settled conviction that the Council now have a great opportunity of contributing to the national benefit through the further development or the chemical industry . . . unless the close relationship between pure scientific research and its technical application be established, the development of our national and imperial resources, so urgently needed at the present time, will be seriously endangered.[2]

In his *Memoirs* Robinson recalls that it was at Liverpool that he learned how to set a proper value on his professional advice. The Harbour Board had been troubled by a series of small fires, arising from oil floating in the docks: Robinson diagnosed the cause and prescribed a remedy. In due course he was asked about remuneration, and diffidently suggested that the Board might care to make a contribution to the University Library. After listening to the comments of his commercially minded friends he concluded that 'This was the first and last time that I made a mistake of that kind.'

He also did consultancy work for Crosfields, the soap manufacturers, whose technical director was E. F. Armstrong—son of Professor H. E. Armstrong, who had a distinguished career in the Central Technical College, London, incorporated in Imperial College in 1911. They visited him at his home near Warrington, where he cultivated a fine rock garden, an activity in which the Robinsons had a lifelong interest. During the war Robinson was a member of the Oils and Fats Committee of the DSIR, and carried out some research on the hydrogenation of linseed oil.

While these contacts with industry provided useful experience outside the rather narrow academic world, and could be a source of useful consultancy fees, few would have supposed that at this stage of his career Robinson had in mind anything but a university career. It is, therefore, a matter for some surprise to find him resigning his chair at Liverpool at the end of 1919 to take up a full-time industrial appointment as Director of Research in the British Dyestuffs Corporation. To understand how this came about, and why the appointment was short-lived, we must digress briefly, and explore a complex, and far from edifying, chapter in the history of the British chemical industry.

Although the synthetic dyestuffs industry had originated in Britain, with W. H. Perkin's discovery of mauveine in 1856, the lead had passed decisively to Germany by the turn of the century. On the outbreak of war Britain was so dependent on German imports that it was difficult even to provide the khaki dye needed for army uniforms. In 1915, the Board of Trade launched British Dyes Limited, having a proposed capital of £2 million, largely government provided, with the intention of bringing all the major dyestuffs companies under one management. In principle, this was a sensible proposal, but it was mismanaged from the start, partly, no doubt, because this was the first attempt by any British government to float a public company. This in itself damned it in the eyes of many, for it undermined the sacred British principle of free trade. Users of dyestuffs very much disliked a condition that customers should undertake to buy their supplies from the new company for a period of time after the end of the war. In this, of course, the government was concerned to protect the industry against a possible resurgence of German dyestuffs. Other difficulties were that nobody with any direct knowledge of the industry was to have a place on the Board, and only a trivial sum was to be allocated for research. These issues were twice debated in Parliament, but despite widely expressed misgivings British Dyes was formally incorporated on 1 April 1915.[3]

The new company failed signally in its declared intentions: worse, it allowed itself to be deflected into activities—such as the manufacture of high explosives and oleum—which were never intended, though the pressure from the Ministry of Munitions must have been enormous. Within a short time it acquired one major dyestuffs company, Read Holliday, but it had no success with its major rival, Levinsteins, nor with some smaller concerns such as Claus and Co. and British Alizarine. Ironically, it was Levinsteins, the outsider, who had valued themselves at only £300 000 in 1914, who made the running, acquiring a synthetic indigo plant at Ellesmere Port in 1916, and quietly buying up Claus in the same year. By that time they were manufacturing about two-thirds of all the synthetic dyes made in Britain, then about 11 000 tonnes annually. The American chemical giant Du Pont acquired an interest at Ellesmere Port, and Nobel Explosives acquired a large enough interest for Harry McGowan—the chief architect of ICI in 1926—to be nominated a member of the Levinstein Board. The whole situation was complicated by personal antipathy between Ivan Levinstein—then a sick man—and James Falconer, chairman of British Dyes.

Such a situation naturally evoked much public criticism, and in 1918 the then President of the Board of Trade, Sir Albert Stanley (later Lord Ashfield), took decisive action. In a public address in June 1918 he was severely critical of British Dyes for failing to do what it had been set up to do: thus encouraged, the shareholders forcibly expressed their lack of confidence in the Board, and the majority, including Falconer, resigned. With the Levinstein star clearly in the ascendancy there was no longer any insurmountable obstacle to the amalgamation sought some four years earlier, though the terms of business took some months to agree. The summer of 1919 saw the creation of the British Dyestuffs Corporation, capitalized at £10 million. It was a holding company, with British Dyes and Levinsteins its main subsidiaries, each, as it may be supposed, wary of the other.

These acrimonious negotiations, and the undercurrents of personal antipathies, were a matter of public knowledge, most particularly so in the Liverpool/Manchester area. Moreover, Robinson would have had direct knowledge of the situation through a connection with British Dyes that had been established in 1916. Conscious of the need for research, but with only limited facilities of its own, British Dyes had engaged W. H. Perkin Jr. to establish and supervise small groups, or 'colonies', in appropriate university departments: each was to be provided with a small number of chemists, at a stipend of £200 a year. Not surprisingly, one of those so engaged was Robinson: other colonies were set up in Cambridge and Leeds. However, the scheme was makeshift, and seems to have achieved rather little. Its progress at Liverpool is briefly recorded in the Annual Reports of the University—the passages in question presumably being supplied by Robinson.

The Report for 1916 records, 'with great satisfaction' that close contact had been established with British Dyes, and 'a branch of the research departments of this firm is, in fact, installed in the laboratory': A. W. Fyfe is named as the first colonist. The 1917 Report records no more than that the association 'has been rewarded by a considerable measure of success'. By 1918 the scale of the collaboration has clearly increased:

> The British Dyes Limited Research Colony now includes four workers. Arrangements have been made for closer co-operation with the factory activities in Huddersfield, and periodic visits are made by each member of the colony to the works.

The reference to Huddersfield is to a new dyestuffs factory that had been built nearby at Dalton.

In 1919, however, university departments generally were gradually returning to normality, though the large influx of ex-service men imposed heavy new demands. We learn from the Report that H. G. Crabtree, one of the colonists, had left on appointment as Lecturer in Colour Chemistry at Leeds. Evidently he had not devoted himself wholly to British Dyes, for he and Robinson jointly published a paper on brazilin chemistry in 1922.

By contrast, Levinsteins' research organization was far more efficient. They had outbidden British Dyes for the services of A. G. Green, who, after long practical experience of the dyestuffs industry in Britain, France, Germany, and the USA, had been appointed Professor of Applied Chemistry (Dyestuffs) at Leeds: his quality had been recognized by his election to Fellowship of the Royal Society in 1915. Under him a group of 30 research chemists had been set up, with good laboratory facilities.

On 21 October 1919 Robinson formally resigned from his chair, as from 1 January 1920, in order to take up an appointment as Director of Research in British Dyes at their Dalton works. This the Council of the University—of which he was a member—'received with regret': also, it must be supposed, with surprise, for on the face of it he was well set where he was. Relations within the department were good, plans were under consideration for expanding the chemical laboratory accommodation and facilities, and the Liverpool/Manchester ambience was congenial. He had ample opportunities for involvement in industrial chemistry without becoming wholly committed to it. He had been only five years at Liverpool, and while clearly ambitious for further advancement he can have had no sense of being in a rut: at thirty-four years of age the greater part of his professional life still lay ahead of him. Why, then, abandon the freedom of academic life—with excellent prospects ahead—for the inevitably more restrictive life of industry, and in particular, plunge into the notoriously murky waters of the British Dyestuffs Corporation?

After this length of time it is still difficult to discern his motives, and his own account in the *Memoirs* is brief, and to some extent contradictory. From this, it appears that in the summer of 1919 Herbert Levinstein, who had succeeded his father as head of the firm on his death in 1916, visited Robinson in Liverpool in his capacity as Joint Managing Director of the British Dyestuffs Corporation, and made a formal proposition to him. We are not to suppose that this came as a bolt from the blue, for the two men were already well acquainted, both professionally and personally. Like his father, who had founded the business in 1864, Herbert Levinstein was a capable chemist, a graduate of the Universities of Manchester and Zurich. Both liked the outdoor life: they had a common interest in mountaineering—though they seem never to have climbed together—and Levinstein was a keen horseman until a bad fall put a stop to this form of exercise. They were clearly good friends.

> I was extremely fond of him in spite of his exploits of derring-do [horsemanship]. His home was at Ford Bank, Didsbury, a very large establishment which we frequently visited.[4]

Though Herbert Levinstein was of German descent—Ivan had been born in Charlottenburg, and studied at Berlin University and Technical High School—he was intent on wresting from Germany her ascendancy in dyestuffs manufacture. No doubt he argued this aspect to Robinson, who states that one reason for his accepting the invitation, after seeking the advice of friends, was that 'it seemed very much in the national interest that I should do so.'[5] Without for one moment questioning his patriotism, this reads a little unconvincingly. The war was over: Germany was crushed, even if potentially resurgent, and citizens generally could feel free to resume their peacetime careers with an easy conscience. Moreover, Robinson was quite sufficient of a realist to know that whatever disadvantaged the German industry was of direct commercial benefit to Levinsteins as a member of the British Dyestuffs Corporation. The precise terms of his appointment are not clear, but in a subsequent application for the Chair at St Andrews he refers to 'very favourable conditions'. How he weighed the pros and cons, and what Gertrude's views may have been, we do not know. What we do know is that the invitation was accepted, and that the experience was not a happy one.

The spring of 1920 saw the Robinsons installed in Fairholme, a house in Mountjoy Road, Huddersfield, not far from the works. There he was provided with a good research staff and a personal assistant, Wilfred Lawson, a Liverpool graduate. As was to be expected, the bulk of the work was directed to the preparation of basic intermediates for existing products and to the development of new dyes—especially azo dyes for the cotton industry. Although not directly related to the natural products which had long been his particular concern, the research was intrinsically interesting, and in retrospect he seems to have regarded it as a worthwhile experience. It certainly gave him a familiarity with dyestuffs chemistry—far greater than Dixon had generously attributed to him when supporting his application for the Sydney chair—which stood him in good stead in later years when he was a consultant to ICI. While it is doubtful whether this sort of work would have satisfied him for long, it was in fact the internal tensions and frustration that made his stay at Huddersfield a brief one.

The union of Levinsteins and British Dyes to form the British Dyestuffs Corporation had been very much a shotgun marriage, and the partners had no liking for each other: relations, as one observer put it, were 'the reverse of friendly'. There were two government-appointed directors, Lord Moulton and Sir Henry Birchenough, and two joint managing directors, Joseph Turner, representing British Dyes, and Herbert Levinstein. The two latter quarrelled incessantly, and Robinson could not disassociate himself: 'They each confided in me their intention to stand on the other's neck'.[6] Additionally, he found himself not fully involved in the Corporation's research programme. At Huddersfield there was already a technical laboratory directed by O. Segaller, to whom problems were first directed: Robinson and his team 'got only some crumbs from the diners' table'. Again, contact with Levinstein's laboratory, where Green had a first-rate team, was 'not the closest that could be desired'.

The bitter quarrels between the two managing directors were not only distasteful to all concerned, but inimical to the Corporation's interests. Towards the end of 1920 Sir William Alexander—a

Brigadier-General who before the war had been associated with Sir Charles Tennant and Company, chemical merchants in Glasgow—was called in to put the house in order. Within less than a month of his appointment he submitted a devastating report, at the end of which he succinctly diagnosed the trouble as 'no system, no co-operation, no organization, no efficiency'. On 18 February 1921 Turner and Levinstein were summarily dismissed from their appointments. Ironically, both nevertheless remained members of the Board: their knowledge and experience of the industry were too great to be dispensed with altogether.

While the appointment at Huddersfield thus had its tribulations, it also brought rewards. It was there, on 10 February 1921, that his daughter Marion was born—a solace for the short-lived daughter born in Sydney—and it was there that in May 1920 he learnt he had been elected to the much-coveted Fellowship of the Royal Society. Shortly afterwards, he was elected to the Council of the Chemical Society, and was thus an acknowledged member of the chemical establishment.

In the circumstances, it is not surprising that his thoughts turned increasingly to returning to academic life, but in the circumstances of the day this was more easily said than done. Having already held two chairs he could scarcely accept an appointment at a lower level; but professorships were then far fewer than they are today, and new ones, such a the one to which he had been appointed at Liverpool, were rare. Once appointed, professors mostly sat tight until obliged to retire. Short of untimely death, or some unexpected quirk of fortune, it was possible to predict which chairs would become vacant, and when. As he considered the prospect in 1921, he must have felt that the prospects for a return to academe were not particularly promising. Dixon was approaching retirement at Manchester, but he was a physical and inorganic chemist, and it was likely that the University would appoint a successor of the same persuasion. At fifty, Lapworth had a dozen years ahead of him as Professor of Organic Chemistry. As we shall see, this is not how things worked out, but so it seemed in 1921.

As it proved, however, luck was with him. At the University of St Andrews, the Professor of Chemistry, J. C. Irvine, had been appointed Principal at the early age of forty-four. As his new duties were too onerous to be combined with those of his chair, the latter fell vacant. Learning of this, Robinson, who knew Irvine very well, wrote to inquire whether he might be considered for the appointment. Irvine sent an encouraging reply, but pointed out that Robinson would have to make a formal application when the post came to be advertised.

He duly submitted his application, set in print as was still the style, in May 1921. It is, of course, essentially a curriculum vitae, but a few points are of interest. The Professor was also to be Director of the Chemistry Research Laboratory, which was concerned with all branches of chemistry. By this time Robinson was a specialist in organic chemistry, but he was at pains to point out that he 'had tried to keep in touch with developments in other branches of the science'. His problems at Huddersfield he glosses over: 'I am not at home in this atmosphere for various reasons which need not be particularised.' No doubt those who had to consider the application were well aware of the situation. However, he gets what mileage he can from his Huddersfield appointment: 'I have thus supplemented my experience and greatly increased my knowledge of the potential applications of chemistry'.

His application was intrinsically a strong one, particularly in view of his recent election to the Royal Society and his impressive list of publications, already more than sixty. He could offer five powerful supporters, headed by W. H. Perkin Jr from Oxford. Two others were Jocelyn Thorpe and F. S. Kipping, who also had supported his application to Liverpool. Finally, there was H. E. Armstrong, retired from Imperial College but still influential, and Sir William Pope, Professor of Chemistry at Cambridge. Interestingly, Pope had recently been approached about the possibility of becoming Vice-Chancellor of Birmingham University, but felt obliged to decline because he had just received a donation of £200 000 from the petroleum industry to extend and equip his department at Cambridge. Had Pope accepted, Robinson might well have been appointed to succeed him; and, as the Cambridge laboratory became the best equipped in the country, Robinson would probably never have moved elsewhere.

His application was successful, and in the autumn of 1921 he and Gertrude—now with their baby daughter—moved on to St Andrews. He was fortunate in the sale of his house to somebody who, for some reason, had a sentimental affection for it, and paid the asking price with no bargaining at all.

5
St Andrews: the Scottish experience

While the plight of the Robinsons in Huddersfield was not such that they were obliged to find any port in a storm, so quickly to find an agreeable haven in the small university town of St Andrews was a stroke of good fortune. Moreover, it had been attained with minimal trouble, for he had been appointed without any personal interview, and after merely the perusal of applications.[1] The milieu was very different from any they had previously experienced in Sydney or in the thriving industrial cities of Manchester and Liverpool. In the 1920s the population of St Andrews, perched on the cliffs of Fife some fifty miles by road from Edinburgh, was less than 10 000: even today it is only 12 000. The university, founded in 1410, is the oldest in Scotland: the other ancient foundations are Glasgow (1451), Aberdeen (1495), and Edinburgh (1583). In 1881 the University College of Dundee was founded some dozen miles away, and in 1897 this had been affiliated to the University of St Andrews: technically, Robinson held office in both.

Despite its antiquity, the University's serious interest in chemistry was then still comparatively recent, dating from the appointment of Thomas Purdie as Professor of Chemistry in 1884. As a young man, he had studied under Johannes Wislicenus in Würzburg, and under his guidance developed a keen interest in stereochemistry—the way in which the atoms of molecules are arranged three-dimensionally in space. In pursuit of this he developed Purdie's 'silver oxide reaction' for methylating aliphatic hydroxyl groups. To the non-chemist this may sound only mildly exciting, but it prompted the great German chemist A. W. von Hofmann to remark:

> Had the St Andrew's School done no more than discover the silver oxide reaction, its name would still deserve to be inscribed in golden letters in the records of chemistry.

In 1895 J. C. Irvine, whose professorship Robinson was assuming, came to St Andrews from Glasgow to work under Purdie, who four years later encouraged him to spend a year with his own old master, Wislicenus, in Leipzig. While there Irvine conceived the idea of using Purdie's reagent to investigate the structure of carbohydrates, which are rich in hydroxyl groups. Purdie was impressed, and invited him back to St Andrews—at his own expense—so that they could work together. Important results were quickly obtained, and a new chapter was opened in the difficult chemistry of carbohydrates—an important group of natural products which includes sugars and starches. Realizing the great prospects, Purdie thereafter made this the main activity of his department, and in 1905 used an unexpected inheritance to build an urgently needed new research laboratory for the University. Many of his colleagues—notably the Nobel Laureate W. N. Haworth and E. L. Hirst—went on to distinguished careers in their own right. Haworth was well known to Robinson, for they had been fellow students at Manchester. Unfortunately, a bad relationship between Irvine and Haworth became notorious. The international reputation of the laboratory ensured a steady stream of first-class research workers: all told, there were about twenty. Emil Fischer, the *Altmeister* of sugar chemistry in Berlin, was pleased to collaborate, being anxious 'to avoid any possibility of collision or competition.'

When Purdie retired in 1909 Irvine was his natural successor. As elsewhere, the First World War made it necessary to divert the laboratories' facilities from research to the preparation of urgently needed chemicals—particularly, rare sugars such as inulin, fructose, mannitol, and dulcitol, needed for medical purposes. By 1921, however, things were pretty well back to normal—though there were still some ex-servicemen completing their studies, which led to a little overcrowding—and thus Robinson took over the direction of a flourishing department with an international reputation: Irvine was a chemist at heart, and tried to maintain his research interests in a private laboratory, but his other duties left him less and less time for this. Nevertheless, he did publish a number of papers up to 1938. In this work Robinson took no part, continuing to work largely on alkaloids. However, sugars were by no means outside his catholic sphere of interest: twenty-five years later, in his last years at Oxford, he insisted on giving the lectures on the stereochemistry of sugars in the general course. Although she now had a small child to care for, Gertrude regularly worked several hours daily in the laboratory, and was registered as a research student. In addition to experimental work, he gave much thought to developing an electronic theory of organic reactions, to which we will come later.

Apart from the prestige of his department and the good facilities in a comparatively new laboratory, St Andrews had much to commend it. The University dominated the small town, and the scarlet gowns of the students were much in evidence. Like Oxford and Cambridge it had developed along collegiate lines. It had a tradition of electing as Rector some well known public figure, not necessarily from the academic world. In Robinson's time it was Sir James Barrie. On the occasion of his Rectorial Address in 1922, honorary degrees were conferred on Ellen Terry, Earl Haig, and John Galsworthy. St Andrews was remote, but not isolated.

The laboratory looked out across the cliffs and the North Sea, and the splendid view more than compensated for the biting winds that often blew off it in winter. Additionally, it was an excellent centre from which to start on walking and climbing expeditions in the Highlands. An obvious disadvantage, however, was its distance—some 450 miles—from London with its twin attractions of the Royal Society and the Chemical Society, whose meetings and other activities were professionally important. There was also the Society of Chemical Industry, catering for the professional interests of industrial chemists. Of all three he was in the fullness of time to be President, as also of the British Association for the Advancement of Science.

Although this was Robinson's third professorship, we must remember that we are still dealing with events of nearly seventy years ago. Direct personal recollections are thus difficult to come by, but Dr James Craik—who was a research student in 1922/3, though not under Robinson—has been able to contribute something.[2] His recollection is that Robinson was happy at St Andrews: he regularly cycled from his home to the laboratory, about three-quarters of a mile away, whistling cheerfully. By this time Marion was attending an infant school. Unlike Irvine he did not lecture to the first-year students, but he took a keen interest in student affairs. In particular, he and Gertrude regularly attended meetings of the University Chemical Society, to which visiting speakers were invited: while waiting for the proceedings to begin, sitting in the front row, they would pass the time with a pocket chess set. He also makes a comment which older generations of chemists, brought up with much less sophisticated apparatus than now prevails, will appreciate:

He must have had a very large and strong hand. It was quite common for him, in demonstrating at a lecture or in the lab., to lift a Winchester [½-gallon bottle] in his right hand and pour a few drops from it into a test-tube held in his left hand.

However, Robinson's stay at St Andrews was destined to be brief, for there had been unexpected developments at Manchester. There, H. B. Dixon's long reign as Sir Samuel Hall Professor of Chemistry—he had succeeded Roscoe in 1886—come to an end with his retirement in 1922. In a surprise move Arthur Lapworth, Professor of Organic Chemistry since 1913, was appointed to fill the vacancy. As a consequence the Chair of Organic Chemistry fell vacant, and Robinson needed no persuasion to accept it when it was offered.

These are the facts of the matter, but it would be naive to suppose that such a felicitous rearrangement took place without considerable contriving, the details of which are not on record. We may suppose, however, that the leading spirit was Lapworth, whose devotion to Robinson, fourteen years his junior, made him ready to make great sacrifices to secure the latter's return to Manchester. Not only had he to adjust to a new field of chemistry in which he was certainly fully competent, but which was not his primary interest—though in modern terminology he can properly be called a physical organic chemist—but also, as Head of Department, he had to take on a great deal of administrative work which would certainly be unwelcome. But the benefits on both sides were considerable. Lapworth once again had daily contact with Robinson, and thus the opportunity to develop their ideas on the theory of organic reactions—and to indulge more readily in mountaineering activities. Robinson now had a much larger department—housed in the relatively new Morley Laboratory—and was much nearer to the centre of gravity of British science than at St Andrews: many Mancunians would say, *at* the centre. The history of chemistry at Manchester has been recorded with meticulous care by G. N. Burkhardt, one of the new colleagues Robinson was to join there.[3] Much useful information has also been recorded by Wesley Cocker, lately professor of chemistry in Trinity College, Dublin.[4]

For their part, the University made Robinson's path easy. Although his credentials were closely examined by a powerful internal committee, there was no advertisement, and thus no alternative candidate. They wanted Robinson, and within the limits of proper academic procedure they made sure that they got him. He formally resigned his appointment at St Andrews and Dundee as from 1 April 1923, to return once again, as professor, to the university where his chemical career had begun twenty-one years earlier.

6

Manchester and London

When Robinson returned to Manchester he was just thirty-six years old, yet he already had three major academic appointments behind him, plus a senior—if short-lived—one in the chemical industry. By any standards this can fairly be described as a meteoric rise, but the more so when it is recalled that before the Second World War professors were commonly appointed much older than they are now. It is, therefore, timely to pause for a moment, and consider the qualities that had so firmly established his reputation that he was actively, and exclusively, sought for perhaps the most prestigious chair of organic chemistry in Britain.

First and foremost, it was the general consensus that he was a chemical genius, a status more easily accorded than defined. Briefly, it may be considered as professional mastery of his subject allied with strong intuitive powers: there is no doubt that his thought processes often by-passed the normal processes of systematic cerebration. If several possibilities were open he had a remarkable facility for choosing the correct one first. We may fairly pay to Robinson a tribute he once paid to W. H. Perkin Jr.:

> He had a wonderful, uncanny instinct for the selection of the best methods for the separation of the constituents of tars and such like unpromising material.[1]

Allied with this, and doubtless to some extent integral with it, he had a memory which was not only extraordinarily retentive, but on which he could draw at a moment's notice. In his later years he could instantly recall—and quite unassumingly, as though it were the most natural thing in the world—technical details of a complex synthesis published decades earlier. Finally, there is ample evidence that he thought pictorially—he could see in his mind the molecules undergoing the transformations in which he was interested. This helps to explain also his skill at chess. In this we may perhaps see an analogy with Rutherford, the pioneer of atomic physics, who was in fact professor at Manchester while Robinson was a student. Once, when dining in the Athenaeum, a fellow member remarked that nobody had in fact seen an atom. Rutherford laid down his knife and fork and said firmly: 'Nonsense! I see the little beggars as clearly as I see this plate of food in front of me.' Robinson had a great regard for Rutherford, and put him, with Louis Pasteur, at the head of his list of scientific heroes.

But of course, the quality of genius is not in itself sufficient for the requirement of a university chair, or at least not for one involving not only guidance of a research group but a fair measure of teaching and administration. In this respect he was, as we have seen, somewhat unpredictable. His lectures might be disappointing to the less gifted students, who in the nature of things would prefer something in the nature of a verbalized textbook, but were very stimulating for the abler ones.

Although his research interest was primarily in natural products, he had acquired—especially through his contacts with the chemical industry—an impressive knowledge of the broad field of synthetic organic chemistry. He took an active interest in students' laboratory work. Yet he was in no sense a narrow academic, for he had a wealth of outside interests: most particularly in music, chess, gardening, photography, and mountaineering. Socially, however, he had one defect which he never wholly overcame. While at times he could be most agreeable and amusing company, at others he could be aloof and brusque to the point of seeming rudeness. Those who knew him well realized that intense preoccupation with a problem currently engaging his interest could make him virtually oblivious of his surroundings, whatever they might be. L. E. Sutton, also a chemist and Fellow of Magdalen, put this point well in a memorial address in 1975:

> A conversation with him could be disconcerting, for he rarely looked one in the eye, but rather tended to look past one as if he were visualizing on the wall behind the structure of some substance in which he was newly interested; but he could also be jovial and great fun.

Even for his friends, such abstractions could be trying: for those who did not understand the situation, or found it unacceptable if they did, the effect could be very unfortunate and long-lasting. This could be aggravated by a characteristic trait of brooking no argument about the chemical conclusions to which his abstractions led him. In fact, there generally was no counter-argument, for he was rarely wrong; but a less dogmatic response would have been diplomatic, and discounted accusations of arrogance.

Having said this, we may be sure that Robinson's estimate of his own capacity was by then such that the invitation to Manchester would have occasioned him no surprise. This is, however, far from saying that he was not extremely conscious of the high status and responsibility of his new office. His immediate predecessor, Lapworth, who had so unselfishly masterminded the appointment, was a valued collaborator and close friend, but for W. H. Perkin Jr., founder of the Manchester school and his own first mentor, he had a particular admiration. After Perkin's death in 1929, Robinson paid warm tribute to him in *The Life and Work of Professor William Henry Perkin*, published by the Chemical Society in 1932. In his seventieth year he went out of his way to pay tribute to the Perkin family of chemists in general and W. H. Perkin Jr. in particular.[2] This included an elaborate chart (here reproduced) 'setting out the academic association of some organic chemists, primarily designed to illustrate the far-reaching influence of W. H. Perkin jun.'. Many of those featured in this chart have already been mentioned, and others will be introduced later. Unfortunately, some readers misread this carefully worded caption, and understood the chart to represent Robinson's assessment of organic chemists: on this erroneous basis, some whose names did not appear were considerably piqued.

To preserve continuity of work in hand, professors commonly take with them some of their current research associates to a new appointment—a mutually satisfactory arrangement. Hitherto, however, this has been outside Robinson's experience—save, of course, for Gertrude. They had returned alone from Sydney, and, understandably, nobody from Liverpool had sought to go with them to Huddersfield: nor, again, had anybody from Huddersfield gone with them to St Andrews. Now, however, the situation was different: from there he brought several research workers to join the very capable team of organic chemists awaiting him at Manchester. One was J. M. Gulland, who went on to a distinguished career via Oxford, The Lister Institute, and—as Jesse Boot Professor of Chemistry—Nottingham. Another was D. D. Pratt, who eventually became Director of the National Chemical Laboratory. He had been Robinson's chief collaborator in an intensive investigation of a group of natural pigments known as the anthocyanins: these are widely distributed in nature in the petals of flowers.

At Manchester it was, of course, Lapworth who was his closest associate: they immediately resumed their development of the theory of organic chemical reactions, which was—as we shall see in a later chapter—to involve them in fierce controversy with C. K. Ingold. Among the junior staff was G. N. Burkhardt, appointed Demonstrator in Chemistry in 1921, and destined for a long and distinguished career in the department. He had a particular interest in reaction mechanisms, and in his later years compiled, for private circulation, a historical work of great interest—*Arthur Lapworth and Others* (1980). Robinson had no difficulty in recruiting able new collaborators. One was Wilson Baker, who joined the department as assistant lecturer in 1924, and left in 1927 to go to Oxford: there they were later to be colleagues again for many years, until Baker was appointed to a chair in Bristol in 1945. Also in 1924 came Alexander Robertson, who collaborated closely in the anthocyanin research: he left in 1928, at the same time as Robinson, to take up an appointment as Reader in Chemistry at Queen Mary College, London: eventually he ended up, in 1933, in Robinson's former appointment as Heath Harrison Professor at Liverpool, as successor to Ian Heilbron. Robertson's subsequent career left Robinson with mixed feelings. In 1957 he retired from chemistry altogether and devoted himself to farming, ultimately owning some 6000 acres in Lincolnshire. Rather ambiguously, Robinson concluded that as his 'chemical research had reached a stage at which no really outstanding developments could be expected with any confidence'—that is, the vein was worked out—'it was reasonable that his farming instinct, inherited from his father, should be indulged'.[3] Yet another of his Manchester colleagues whom he particularly recalls in his *Memoirs* was W. Bradley, who also participated in the anthocyanin work. He, too, went on to Oxford, at the same time as Robinson, and eventually became Professor of Colour Chemistry at Leeds.

Like other keen chess players, Robinson enjoyed playing several games at the same time, and he showed the same facility in his research. While he was preparing research papers with his immediate colleagues he was collaborating with others at a distance. Thus 1922 saw the publication of a paper on the alkaloid harmaline jointly with W. H. Perkin Jr. at Oxford and W. O. Kermack, whom he had known at St Andrews as the head of the research laboratory of the

Royal College of Physicians of Edinburgh. Sadly, Kermack was blinded in a laboratory accident in 1924, but nevertheless managed not only to continue in this appointment until 1948 but to serve for another twenty years as Professor of Biological Chemistry at Aberdeen. In 1926 he was publishing jointly with H. R. Ing of Oxford, who later became Professor of Pharmacological Chemistry there.

He was also involved with Arthur George Perkin, brother of W. H. Perkin Jr. Since 1916 Arthur had been Professor of Colour Chemistry at Leeds, succeeding A. G. Green, who, as we have noted earlier, had gone off to direct research for Levinsteins. Save for a common passion for music and organic chemistry the brothers were quite dissimilar: Arthur lacked ambition and had what Robinson obscurely called 'a rather old-fashioned manner': he saw nothing old-fashioned in his experimental skill, however, for he ascribed to him 'one of the most skilful pairs of hands which has served our science'. Robinson acted as his external examiner, a task which he found stimulating but also wearing: Perkin's practice was to read aloud every word of every script! Robinson also knew a third Perkin brother, Frederick Mollwo Perkin. Unlike his brother Arthur he had a very practical turn of mind, and made a successful career as a technical consultant.

With A. G. Perkin, Robinson enjoyed what may be called a chemical *divertissement*. Perkin also was keenly interested in natural products, in his case the pigments derived from the parent molecule anthraquinone: in 1918 he wrote a book on the subject with A. E. Everest. At one stage he acquired from a London curio dealer a dark brown lump consisting of carajura, allegedly used by the Indians of the Upper Orinoco as a cosmetic: it was so highly esteemed that it was widely acceptable there as a form of currency. It seemed to them to derive from some sort of bush rose having violet flowers, and they set about identifying the active principle. By various processes of degradation they extracted a pure substance to which they gave the name carajurin.

In those days, but less so now, structures so deduced had to be confirmed by synthesis—that is to say, the original substance had to be recreated by a series of unequivocal reactions from simpler, well-identified, substances. In the case of carajurin this confirmation was effected by one of Robinson's research students, T. R. Seshadri, who returned to India in 1930 and eventually became Professor of

Chemistry in Delhi University. This exemplifies a continuing feature of Robinson's research—collaboration with a succession of able students from overseas. Also to this phase belong, among others, another Indian student K. Venkataraman; three Japanese, Hidegiro Nishikawa (who was with him in St Andrews), Shinzo Murakami, and J. Shinoda; and R. H. F. Manske, of German origin but trained in Canada, to which he returned, eventually to become director of research in the Union Rubber Company, Guelph. Manske was deeply interested in alkaloids, and acquired a profound knowledge of them: in his later years he published a ten-volume book, *The Alkaloids* (1950–65), which was acknowledged as a standard work. Even birds of passage became brief collaborators. Thus when Professor Y. Asakina, of Tokyo University, visited England in 1926, he took the opportunity of calling on Robinson in Manchester. The result was the synthesis of a substance known as ruteocarpine—a plant-derived drug used in China—duly described in the *Journal of the Chemical Society* in the following year.

Long-term, this had the result that many of his hand-picked students became established in senior posts in many parts of the world. When, in his later years, he travelled extensively, these former research associates were to be found—to his and their pleasure—on many committees of welcome. Nor did he reserve these important contacts for himself. Many people, including myself, enjoyed the benefits of letters of introduction which he readily wrote on our behalf, usually quite spontaneously.

While organic chemistry is a pretty exact science, it is not totally so. A series of reactions may seem capable of more than one interpretation, or other workers may find difficulty in repeating a particular experiment that has been described. Although he kept his main ammunition for Ingold, Robinson had his share of minor skirmishes—indeed, he probably relished them. Thus he went to some trouble to demolish an erroneous view of the structure of anthocyanidins incautiously advanced by Nierenstein of Bristol University: the latter's main research interest was, as it happened, in the chemistry of the very intractable group of substances known as tannins. Similarly, with the collaboration of F. L. Pyman—almost an exact contemporary during his student days at Manchester—he vindicated himself in the face of criticism by the distinguished Swiss chemist, Amé Pictet, of his structure for the poppy alkaloid papaverine. He and Pyman had become near neighbours again, as the latter was head of the department of applied chemistry at the Manchester College of Technology. In 1927 he was appointed Director of the Research Department of Boots Pure Drug Co. at Nottingham, and he and Robinson, as consultant, then worked together on potential synthetic antimalarials and antiseptics. There was at that time widespread interest in a synthetic substitute for quinine. The Germans were the first to succeed, with the introduction of atebrine in 1932. The antimalarial work had an amusing outcome. It was known that a derivative of the alkaloid harmine had some antimalarial activity, so between them they made a succession of analogous derivatives with very bulky side-chains attached to the parent harmine molecule. This seemed promising—until it was discovered that the antimalarial activity, 'not of the most striking order', resided in the side chain alone, the harmine by then being swamped and making no contribution. Robinson's references to Pyman in his *Memoirs* reveal another facet of his talent. Whenever they met at Nottingham, they always found time for a game of billiards or snooker after lunch at the County Club. Robinson could hold his own, he relates, at billiards, but was quite outclassed at snooker. Clearly, Robinson not only liked Pyman as a friend, but also had a high regard for his chemical ability, 'especially the brilliant synthesis of histidine'. He lamented the failure of both the Royal Society and the Chemical Society fully to recognize his merit—he was elected FRS in 1922, but received no other recognition—and puts this down to 'his unassuming manner and complete lack of self-advertisement'.

This second period at Manchester must, for the most part, have been a happy one for the Robinsons. The research on alkaloids and plant pigments was going well and widely acclaimed, and he and Lapworth were sharpening up their ideas on the theory of reaction. They had a comfortable home and a wide circle of agreeable friends, though during this time Gertrude's father died. Manchester and its environs provided ample occupation for their leisure, whether indoors or out. Sadly, the end of their stay was clouded. At the end of 1925 Gertrude again became pregnant and a son, Michael Oliver, was born on 4 September 1926. It gradually became clear, however that he was handicapped and a victim of Down's syndrome, an incurable condition needing special and devoted care. However, the nature of the disease was not then understood, and the Robinsons spent a great deal of money with Swiss and other specialists in a vain search for a cure. In those days mental handicap was less sympathetically regarded than it is now, and it is very much to his parents' credit that, as Michael grew older, they made no bones about taking him to public places such as restaurants.

Possibly it was the unsettling effect of this that led to a new and rather puzzling move on Robinson's part: he accepted an invitation to become Professor of Organic Chemistry in University College, London. His own account is unilluminating:

I accepted the invitation to occupy this Chair with some hesitation, since I was so happy in my activities at Manchester, but it was a kind of usual sequence which had been followed by others, for example, by A. V. Hill, the physiologist.[4]

Hill was a Cambridge graduate, who had been Professor of Physiology at Manchester (1920–23), where he had both resuscitated a run-down department and carried out brilliant research on the physiology of muscular and nervous activity. He and Otto Meyerhof shared the 1922 Nobel Price for Physiology or Medicine. Robinson and Hill must have known each other at Manchester during a short overlap, but there is no indication of a particularly close relationship at that time. So far as they were in contact subsequently, we know that Hill would certainly have spoken favourably of University College and the pleasures of living nearby in north London.

No doubt an additional factor was that University College, founded in 1828 as the first constituent of the University of London—by which title it was originally known—was traditionally non-sectarian. This was in contrast to King's College in the Strand, founded a year later under Anglican auspices. Thus Robinson would find himself in the kind of liberal milieu to which he had been accustomed.

A search of the College records[5] throws no light on the matter. The Annual Reports do no more than formally report his appointment in 1928, and, in 1930, his resignation to succeed W. H. Perkin Jr. at Oxford. Nowhere is there any evidence of a rift at Manchester. What Lapworth's feelings may have been we do not know; certainly he must have been disappointed that the man whom he had made such sacrifices to attract had left after so relatively brief a stay. A moving tribute to Lapworth which Robinson undertook after his death contains some interesting comment in this respect:

As administrator of his department and laboratory he showed firmness and wisdom, but this was not his real *métier* and it is a pity that a sense of duty compelled him to undertake responsibilities which turned out to last for too long . . . Lapworth's health, at least in later years, was never robust, and the exceptional strain of carrying the burden of the whole Department of Chemistry eventually broke it down.[6]

This duty and burden had, of course, been assumed solely in order to make Robinson's appointment possible, so it is arguable that he had an obligation to continue in it unless he had good reason to leave. However, it may well be that Lapworth urged him to accept the London appointment in furtherance of his career. In any event, however, Robinson would not have continued at Manchester after 1930, when he was invited to the prestigious Waynflete Chair at Oxford in succession to Perkin: none would have expected him to refuse this.

Whatever the circumstances, the move to London seems to have been congenial. The invitation to go there was a consequence of the retirement of J. N. Collie, an accomplished and versatile chemist who had built up a fine school of organic chemistry since his appointment as professor in 1902. This followed an internal reorganization in which Sir William Ramsay, then already world famous for his research on the inert gases of the atmosphere, and later as a Nobel Laureate (1904), had assumed a new title as director of the chemical laboratories. Collie was a man of wide interests, ranging from incunabula to Chinese porcelain, from Japanese carving to vintage clarets. He was also an accomplished painter in water-colours. But he was also a man of practical ability: in particular, in the laboratory he was a very skilled glassblower. Robinson knew him already, and clearly liked him; which was just as well, because Collie carried on in the College as Emeritus Professor and Honorary Lecturer: he was also provided with his own research laboratory. But the greatest bond between them must surely have been mountaineering, to which Collie's devotion was certainly no less fanatical than Robinson's. He had a particular affection for the Coolin in Skye, where he frequently spent the long vacation at Sligachan: Sgurr Thormaid (Norman's Peak) is named after him. Robert and Gertrude made many climbs with him there. But his climbing interest was not merely domestic—like the Robinsons, the world was his oyster. He had climbed in the Canadian Rockies, the Alps, and the Lofoten Islands, and as a young man had been a member of the ill-fated 1895 assault on Nanga Parbat in the

Himalaya, in which A. F. Mummery lost his life. Not long before Robinson went to London Collie had been President of the Alpine Club. Like Robinson, he was a keen photographer, and used his photographs to illustrate some thirty articles and two books on mountaineering.

Typically, when the Robinsons made their first appearance at University College, they were themselves fresh from a five-week climbing holiday in the Alps. They had found themselves a house in Hendon, at 44 Brampton Grove, from which it was an easy drive down Finchley Road to the College with—at that time—ample parking space in the quadrangle. Its name was Bramtoco, which they innocently supposed to have some Red Indian connotation. They learnt later that it had formerly belonged to an employee of the British American Tobacco Company.

While the friendship with Collie must have been particularly significant, there were others whom he also recalls with pleasure in his *Memoirs*. One was F. G. Donnan, Professor of General Chemistry, whom he had known at Liverpool. Another was E. N. da C. Andrade, Professor of Physics, with whom he often had lunch. Percy, as he was universally known, was a gifted physicist noted particularly for his discovery of 'creep' as a common cause of failure in metals. Robinson summed up his character very succinctly:

> Andrade's emotional range was wide. At one end of the scale he wrote a book of poems, and at the other his highly irascible temperament got him into serious trouble . . . I found his personality highly stimulating and was always pleased to be in his company.

As at Manchester, Robinson welcomed research workers from abroad. One of these was Holger Erdtman, from Stockholm, who came with a Ramsay Memorial Fellowship. He appealed to Robinson on two counts. First, he was keenly interested in his theory of organic chemical reactions: second, he had worked with natural products. The latter led him to develop a system of the taxonomy of plants, especially conifers, based on the chemical composition of their tissues. This approach certainly revealed some unsuspected relationships between plants, and helped to confirm others; but generally speaking botanists were sceptical. Erdtman was a colourful character: when I first met him in Stockholm, carrying a letter of introduction from Robinson, I was somewhat startled to be greeted by somebody in a purple knickerbocker suit, yellow stockings, and buckle shoes. But I remember him, too, as a kind host and stimulating conversationalist.

Another foreign worker was V. M. Mičovič, who joined him in his continuing research on brazilin: he later became Professor of Organic Chemistry in Belgrade. Other visitors included Seshadri, who had worked with him in Manchester; K. N. Menon, also from India; Gerald Schwarzenbach from Zurich; and S. Sugasawa from Japan.

Close by University College was the London School of Hygiene and Tropical Medicine, where Harold Raistrick was appointed Professor of Biochemistry in 1929. He was interested in the chemistry of micro-organisms, and this interest in natural products was a bond between them; later they were to publish some joint papers.

Rather little of a personal nature is on record for this period of Robinson's life, and it is fortunate that after his death one of his students there, Catherine Tideman, put some personal recollections on paper.[7] In 1931 she married the late Raymond Le Fèvre, appointed lecturer in organic chemistry at University College in 1928. He subsequently became (1946–71) Professor of Chemistry in the University of Sydney, and thus ultimately a successor to Robinson there.

She gives a vivid account of Robinson's lecturing style:

> Robinson's lectures were always conducted with enthusiasm, there was an air of excitement about them. He had the delightful trait of breaking off to give us the pros and cons of some topical chemical controversy, and if it involved his personal research and he felt he had become a little vehement in his utterance he would invariably say, 'I must apologise, I feel very deeply about this.'

Leaving aside these characteristic digressions from his main theme, he had a fairly standard approach. At the top left-hand corner of the blackboard he would write a structural formula A: at the bottom right-hand corner he would write another formula B. The lecture—with practical demonstrations at the front of the class—consisted of discussions of possible synthetic steps from A to B. This technique he carried on into examination papers:candidates had to make their way as best they could from A to B, not always convincingly:

> . . . in due course (we) were summoned to appear before the great man. I can still recall his comment . . . 'I've wanted to to identify you; do you know that you have revolutionised organic chemistry?'

Robinson took an active interest in all his students, regularly touring the benches to give advice and encouragement. He was also President (1929–30) of the Chemical and Physical Society.

Mrs Le Fèvre recalls also the part played by Gertrude in the life of the chemical community:

> Mrs Robinson, as she then was, was most hospitable to all her husband's collaborators, she regularly entertained them in her home . . . they were really friendly warm people who enriched the lives of all who knew them, their time at UC was all too brief for those who had the good fortune to be there with them and my husband and I are proud to be numbered among those fortunate people.

Quite recently (1987) she had a further interesting recollection. At that time the research on the anthocyanin pigments of plants was at its height, and large quantities of vegetable material had to be crushed to release the pigments. The laboratory had no suitable crushing plant, and the Robinson's had to improvise. They put the plants in a heap on the ground, covered them with boards, and drove their car backwards and forwards over them.

However, Robinson's stay in London was destined to be short-lived. In September 1929 Perkin died, and the vacancy was advertised in the Oxford University *Gazette*: the salary was to be £1200 plus £200 as head of the laboratory. Whereas Perkin's appointment seems to have been made on an informal basis, there now existed a formally constituted Board of seven Waynflete Electors. These were *ex officio* the Vice-Chancellor (then the Revd. Frederick Homes Dudden, Master of Pembroke and chaplain to George V) and the President of Magdalen (G. S. Gordon, formerly Professor of English Literature). The other members, appointed on a three-year basis, provided the necessary chemical weight. They were Sir William Pope (Professor of Chemistry at Cambridge); Sir Henry Miers (formerly Waynflete Professor of Mineralogy at Oxford, and, until very recently, Vice-Chancellor of the University of Manchester); N. V. Sidgwick and B. Lambert, representing Oxford chemists; and last, but by no means least, his old friend and colleague Arthur Lapworth. A meeting of the Electors was quickly convened, and Robinson was duly elected; but he was not free to take up the appointment until 1 July 1930, and this created what might have been an awkward situation at University College, London. There Ingold, then Professor of Chemistry at Leeds, was elected to succeed

Robinson: Donnan, as head of the department, is rumoured to have preferred a different appointment.[8] Be that as it may, Ingold duly arrived at University College, and did so some months before Robinson departed. In view of the bitter controversy in which they had been engaged over the mechanism of chemical reactions, and Donnan's supposed opposition to Ingold's appointment, there was clearly a potentially explosive situation. In the event, however, all three got on quite amicably during the brief overlap.

The notorious Robinson/Ingold controversy has already been referred to several times, but for two reasons this seems to be the most appropriate point at which to explain what it was all about. The first is that, although it grumbled on for years, and to almost literally his dying day Robinson spoke bitterly of Ingold, by 1930 the main battles had been fought. The second is that Robinson's appointment to Oxford was very much a turning-point in his life. He himself was conscious of this, for the first volume of his *Memoirs* ended with his London appointment, and the second (never completed) was to have dealt primarily with the Oxford years. To interpolate the Ingold affair at this point is, therefore, as logical as any other choice.

7
The electronic theory of reaction: the Ingold controversy

'The development of these ideas constituted, in the writer's opinion, his most important contribution to knowledge.' So wrote Robinson[1], at the very end of his long life, but it is doubtful if many modern organic chemists would agree with his self-evaluation: he is much better known for his work on natural products. His electronic theory of organic reactions was undoubtedly important in its day, as providing some logical explanation of hitherto largely empirical observations, but he did not adequately publicize his ideas nor update them to match advances on other fronts, and they are now largely of historical interest. It is to be feared that there is more interest in the bitter public controversy between Robinson and C. K. Ingold over matters of priority than over the theory itself. This applies with even more force to general readers of a book such as this, for a critical evaluation of his theory demands a more than elementary knowledge of organic chemistry and the esoteric shorthand inseparable from its presentation. For those able and willing to digest the development of Robinson's electronic theory in all its chemical complexity, a number of authoritative reviews are available, including several expositions of his own.[2–8] In the present context we can attempt no more than a brief and necessarily over-simplified summary of his contributions and how they accorded, or failed to accord, with those of his contemporaries, notably of Ingold.

As a preliminary, we must remind ourselves that when Robinson was formulating his ideas in the 1920s, organic chemistry was far less sophisticated than it is now. To put it in perspective, it was for practical purposes less than half as old as, and correspondingly less mature than, it is today. In 1803 John Dalton had given substance to the philosophical concept of atomism by asserting that atoms were indivisible, and that the different elements were characterized by differences in their atomic weights, which could be ascertained by experiment. Elementary analysis of compounds showed that atoms combined with each other in fixed proportions, and that each had a limited combining power. From this developed, around mid-century, the concept of some sort of saturation value or valency, affirming that for any given atom there was a limit to the number of other atoms that could combine with it. In the context of organic chemistry, then in its infancy, these ideas were developed by F. A. Kekulé, of Heidelberg, in his *Theory of Molecular Structure* in 1858, less than thirty years before Robinson was born. In this he advanced two fundamental postulates, which, though very simple, brought to the confusion of organic chemistry an orderliness which hitherto had been sadly lacking. The first was that carbon was quadrivalent; that is, that it could combine with four univalent atoms such as those of hydrogen; two bivalent atoms such as those of oxygen or sulphur; and so on. The second was that carbon atoms have a remarkable capacity for linking together to form chains. Thus an organic chemical molecule consists of linked carbon atoms, to which are attached a variety of other atoms or groups of atoms (radicals).

Traditionally, but perhaps apocryphally, the germ of this idea came to Kekulé in a dream, as he dozed his way home late at night on the outside of a London omnibus. Thus was born the Structure Theory of organic chemistry, which replaced the earlier Type Theory of C. F. Gerhardt, which was essentially classificatory, and gave no indication of how atoms were arranged relative to each other in the molecules. With one exception, to which we will come later, Kekulé was timid in following up the implications of his new ideas, and it was in the first instance his student, J. W. F. A. von Baeyer (1835–1917) who used the Structure Theory to assign structures to the rapidly growing number of organic chemicals. In due course (1875) von Baeyer was appointed Professor of Chemistry in Munich, and among his students there, in 1882, was W. H. Perkin Jr., destined to be Robinson's mentor at Manchester. Thus when Robinson entered Manchester University as a student in 1902 the organic chemistry he was taught had a direct link with the pioneers of the Structure Theory, on which a considerable edifice had been built.

Although Kekulé largely failed to follow up his own theory, he did nevertheless make one further major contribution. In 1865, by which time he was Professor at Ghent, he again let his mind wander as he dozed by his fireside:

> Again the atoms were gambolling before my eyes. This time the smaller groups kept modestly in the background. My mental eye . . . could now

distinguish larger structures of manifold conformation: long rows, sometimes more closely fitted together; all twining and twisting in snake-like motion. But look! What was that? One of the snakes had seized hold of its own tail, and the form whirled mockingly before my eyes. As if by a flash of lightning I awoke . . . and spent the rest of the night in working out the consequences of the hypothesis.

The significance of his dream was that he had realized that carbon atoms could join together not only in chains but in rings. That he should have made the discovery through this particular vision of snakes is singularly appropriate, for the ancient Greek symbol of the *Ouroboros*, the tail-eating serpent, was much favoured by the alchemists of classical Greece and ancient Egypt.

In modern terminology, this was a major breakthrough, for it meant in effect that all the vast range of organic compounds could be delineated schematically in terms of straight or branched chains or rings of carbon atoms, with a variety of side-chains attached. In some compounds the rings might be fused to form polycyclic molecules, and in others the chains and rings might contain atoms other than carbon, such as nitrogen and sulphur. Nevertheless, out of chaos, order had been brought, in terms of a few simple principles. For Robinson, it engendered a lifelong fascination:

I was fascinated by the beauty of the organic chemical system. Indeed, I am disposed to agree with Sir Frederick Gowland Hopkins, who once declared that the system of organic chemistry is one of the greatest achievements of the human mind.[9]

We can perhaps see some similarity in the mental processes of Kekulé and Robinson. The former conceived some of his greatest achievements in a dreamlike state, and used to advise his students: 'Let us learn to dream and then perhaps we shall learn the truth . . . but let us beware of publishing our dreams before they have been put to the proof of the waking understanding.' We may compare this with the periods of total abstraction for which Robinson was well known: in these, no doubt, he saw the molecular components of an organic chemical reaction as clearly as Kekulé saw the writhing chains of carbon atoms joining head-to-tail. Like Kekulé, Robinson would then put his intuition to experimental test, leading on eventually to a formal paper in, as a rule, the down-to-earth pages of the *Journal of the Chemical Society*.

As we have remarked, Kekulé himself was slow to perceive the implications of his own theory, so it is not surprising that it was not immediately accepted by his contemporaries. Although basically correct, there were undeniably anomalies to be explained. Thus in union with hydrogen, two atoms of quadrivalent carbon should form only one compound C_2H_6 (ethane): in practice they form also C_2H_4 (ethylene) and C_2H_2 (acetylene). A particular difficulty arose in connection with the hexagonal molecule of benzene, C_6H_6. Apart from the fact that quadrivalent carbon required a formula C_6H_{12}, there was no evidence to suggest that the molecule was anything but symmetrical: every carbon atom identical with every other. Yet experiment showed that if one hydrogen atom was replaced by a different atom (or radical) the reactivities of the remaining carbon atoms were no longer identical. Then, in 1874, a new complication was introduced by J. H. van't Hoff, a former student of Kekulé, who postulated that the four valencies of carbon did not lie in a single plane, but were directed to the four apices of a tetrahedron. This, literally, introduced a new dimension to organic chemistry: molecules were not flat, but three-dimensional bodies. This was of more than academic interest, for it required that certain molecules could exist in two forms, which were essentially mirror-images of each other. A simple analogy is the left hand and the right hand of the human body: although for practical purposes the same, they are not in fact superimposable. Living cells can be very sensitive to these subtle differences: thus tartaric acid can exist in two forms, one of which is readily broken down by a *Penicillium* mould, while the other is untouched. In the case of large molecules, such as proteins, there may be a very great number of such structural variations, and often enzymes, which regulate so many of the bodily functions, will react with only one of them. For a chemist such as Robinson, with a particular interest in natural products with large molecules, these stereochemical considerations were very important.

By the turn of the century, structure theory was well established, and organic chemists could go about their business systematically. There would be no repetition of the experiment which brought fame and fortune to the original W. H. Perkin—virtually the founder of the organic chemical industry—in 1856: he then obtained the first synthetic dye, mauve, as a result of a hopelessly misconceived project to synthesize quinine. His son, W. H. Perkin Jr., would not have fallen into such an error—though undeniably his better chemistry made him less rich! However, his approach to chemistry was still essentially pragmatic, which Robinson—despite his immense regard for him—recognized as a serious defect:

He had a simple outlook on the subject, which he treated as a system in which the parts were related by reactions. However, he never paused to consider the underlying mechanism of organic chemical reactions. For him, chemistry was still the science for transformation of materials and these would be represented by changes in structural formulae. Synthetic and other reactions could usually be explained by drawing rings round reactive moieties, with consequent changes in the bonding of bonds.[10]

This empiricism was typical of many chemists of Perkin's generation, but some younger men—Robinson among them—sought a deeper insight, an understanding of the mechanism by which chemical transformations were effected. In this they had a double motivation. The first was intellectual, and natural to a scientist: it promised a significant advance in our understanding of the working of nature. The second was practical: an understanding of how chemical transformations took place at the molecular level held out the possibility of planning syntheses more certainly and more effectively.

Fundamental to the new thinking was the development of ideas about the concept of valency.[11] Kekulé's system was based on an invariably quadrivalent carbon atom; that is, one having four bonds available for the attachment of other atoms, including those of carbon. But there was no idea at all about the nature of these bonds—nor could there have been in the light of the primitive atomic theory of the day. They might have been symptomatic of electrical, magnetic, or gravitational forces, or even some kind of mechanical linkage. All was speculation: all that seemed clear was that they were strong enough to give compounds specific structures under ordinary conditions, but weak enough to be broken in the course of chemical reactions. Kekulé, rather weakly, affirmed that his formulae were only formulae indicative of reaction potential, and not micro-models of the molecules themselves. This was, of course, true to the extent that chemists expressed their formulae two-dimensionally on sheets of paper, whereas the molecules themselves were undoubtedly three-dimensional. Kekulé arbitrarily left the unaccountable valencies as spare and unattached, or—if they were on adjacent carbon atoms—expressed them as double or triple bonds (C=C, C≡C), ignoring the fact that such bonds are points of weakness, in that they are more readily attacked chemically than a single bond. Others gave up the struggle and simply inserted single bonds between all atoms regardless.

A particular point of difficulty arose in formulating benzene (C_6H_6), parent of a vast number of substances important in the manufacture of drugs, explosives, dyes, plastics, and other industrially important substances. On Kekulé's theory it can be convincingly expressed as

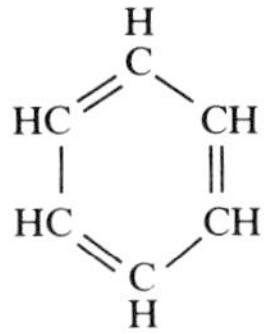

H
C
HC CH
HC CH
C
H

with single and double bonds alternating around the ring. However, if two adjacent hydrogen atoms were to be replaced by, say, chlorine (Cl) atoms, two different compounds would be expected, namely

Cl Cl   Cl Cl

according to whether the chlorine atoms attach themselves to adjacent carbons linked by single or by double bonds. In fact, only one such compound is formed, indicating that all the bonds are equal. Much ingenuity was expended in explaining this, notably a supposition that some kind of internal resonance occurred within the benzene molecule, the single and double bonds alternating so rapidly that, as it were, an approaching chlorine atom had no time to distinguish one from the other.

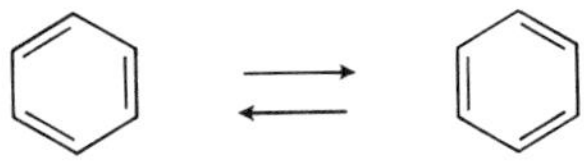

A further difficulty with Kekulé's benzene formula is that it raises expectations that because of its surplus valencies the molecule would be rather particularly reactive, which in fact it is not. This is in marked contrast to butadiene, a straight-chain molecule, in which, again, a single bond is located between two double ones:

$$CH_2 = CH - CH = CH_2$$

Here the two terminal carbons are notably more reactive than the central ones.

Thus the last decade of the nineteenth century saw a situation not uncommon in scientific research generally. A major advance had been made—the Structure Theory—but the new prospects opened up were far from fully explored.

Then, at the very end of the century, a German chemist, F. K. J. Thiele, formulated his theory of partial valencies, supposed to exist on unsaturated carbon atoms. In the case of butadiene the partial valencies of the two central carbon atoms would neutralize each other, leaving free partial valencies available for reaction on the end ones.

$$\begin{array}{ccccccccccccccc} C & = & C & - & C & = & C & \longrightarrow & C & = & C & - & C & = & C \\ \vdots & & \vdots & & \vdots & & \vdots & & \vdots & & & & & & \vdots \end{array}$$

If this notion were extended to benzene, there would be a total internal neutralization, making all the carbon atoms equally reactive.

Thiele put forward his ideas in two papers published in 1899, and they attracted wide interest. The date is significant in the present context, because it was just three years before Robinson entered Manchester University, and these new ideas would have been much discussed among chemists there, staff and students alike. Indeed, Robinson acknowledged that Thiele's theory was the starting point for his own ideas about the course of chemical reactions. That, however, was still far in the future, and for the moment we must go back and recall Robinson's close friendship with Arthur Lapworth at Manchester from 1909 to 1912. Lapworth then already had a strong interest in reaction mechanisms, and had developed his concept of alternating electrical polarities in a chain of carbon atoms, which served to explain how reactivity at one site in a molecule could influence reactivity at another. Embodied in this was the concept of a directing or key-atom which exercised its influence at a distance:

The extension to the influence of the directing or key-atom over a long range seems to require for its fullest display the presence of double bonds and usually in conjugated [alternating] positions; consequently the principle must find ample scope in the aromatic [benzenoid] series where conjugation is the rule.

Robinson eagerly assimilated Lapworth's ideas and contributed others of his own:

I had the inestimable advantage of friendship with him, and unlimited opportunity for discussion, from that time [1909] until 1912, when I entered on my first Professorship at the University of Sydney. During that period he was possessed by his theory of alternate polarities which, though not a reliable guide in all circumstances, still provides a useful mnemonic for the behaviour of molecules of many carbon compounds in reactions.[12]

By 1916, when Robinson had returned to Liverpool from Sydney, he had adapted, with modification, Thiele's theory of partial valency and Lapworth's alternating polarities as the basis of his own ideas.

The starting point was the theory of partial valency which differed from the well known views of J. Thiele in an important respect. Thiele suggested that an unsaturated atom possesses a partial or residual valency in addition to those normally represented by ordinary bonds. My idea was that of bonds themselves being split and thus providing partial valencies. The difference between the two conceptions may appear exiguous but turned out to be important in that the theory of divisible valency could be used to illustrate reaction mechanistic theories which were capable of direct translation on the basis of electronic theory of valency . . . A second stage in the development of my own theoretical ideas was the recognition that when a bond is divided it must *ipso facto* be polarised.[13]

A year later, in 1917, he was satisfied that he had evolved an all-embracing theory of organic chemical reactions, referring to

a system of mechanism of reactions which appears to be capable of including the representation of chemical changes of the most varied type, and the present authors R. R. and G. M. R. are not acquainted with any examples of reactions the course of which cannot be illustrated in the manner implied.[14]

The above reference to the 'electronic theory of valency' demands some explanation. Lapworth's original alternating electrical polarities were essentially conceptual: he did not think of measurable electrical peaks and troughs. Later, however, G. N. Lewis advanced the concept of covalent bonds between atoms, which consisted of pairs of electrons, one derived from each atom, and this was developed by Irving Langmuir. Such bonds gave each atom an outer shell of eight electrons, known as an octet, conforming to the stable configurations of the inert gases. In 1921 Langmuir gave a presentation of this concept at a meeting of the British Association in Edinburgh, at which Robinson was present. He at once saw the implications for his own theory, and reformulated the latter accordingly. In the following year he published, with W. O. Kermack, 'An Explanation of the Property of Induced Polarity of Atoms and an Interpretation of the Theory of Partial Valencies on an Electronic Basis.' This incorporated the concept of the stable octet of electrons, to which he later added what he called the 'aromatic sextet', represented as

to denote a six-electron system that gave benzene its special properties. Later, however, in one of his rare admissions of error, he concluded that

> The 'aromatic sextet' is not a vital part of the theory and severe self-criticism might even suggest that it is a phantasy.[2]

Surprisingly, while Robinson took the new electronic theory of valency pretty much in his stride, he failed to see the relevance of Max Planck's quantum theory, according to which energy is transmitted not continuously but in discrete quanta, much as matter is not infinitely divisible but is ultimately resolved into atomic particles. The circumstances, according to Todd and Cornforth,[6] are curious. In 1931 W. K. F. Hückel published a theoretical paper on the directive influence of substituent groups attached to a benzene molecule which reached conclusions diametrically opposite to those reached by both Robinson and Lapworth on chemical grounds. They published reasoned replies in *Nature*,[15] to which Hückel airily replied that his views were based on quantum theory, and therefore could not be challenged. In the event, it turned out that Hückel had made certain assumptions which were not valid, and was wrong. Characteristically, Robinson decided he would have no more truck with quantum theory, and it was left to others to develop it in the context of reaction mechanisms. Todd and Cornforth noted this with regret:

> Robinson's estrangement from quantum theory was a very sad thing. There can be no doubt that he could have mastered the relevant parts of it not only without difficulty but also without any real effort, had he so wished. Had he done so, theoretical organic chemistry would probably have followed a very different course . . . many of our current advances might have been made a quarter of a century earlier.[6]

This may well be true, but in fact at the time of this brush with Hückel his interest in the theory of reaction—which had occupied him for some twenty years—was declining, though he continued to use it as a working tool. A. J. Birch sees this change as a matter of temperament:

> . . . he had little feeling for physical chemistry. He liked the broad imaginative sweep of ideas and having made a general point to his satisfaction would depart to new creative areas. He felt that he had solved a problem in principle with a general suggestion, and was impatient of the exacting and often boring experimental work then required to develop it experimentally . . . Part of the reason was his wide range of interests and increasing involvement in outside affairs but probably his temperament, which rejected the idea of a dull grinding of figures, was mostly responsible[16]

But for an unexpected circumstance, it may well have been that Robinson would have published no more on the theory of reaction as such, as distinct from referring to it incidentally in papers on other subjects. The unexpected circumstance was that Ingold began to publish a series of papers on reaction mechanisms which Robinson came to regard as claiming credit for the views earlier put forward by himself and Lapworth. It developed into a sustained, and not very edifying polemic, which for a time excited much interest among those standing on the touchlines. To understand what it was all about we must for a moment go back on our tracks.

Although Lapworth and Robinson were undoubtedly pioneers in seeking some systematic explanation of why organic reactions followed the course they did, they did not have the field to themselves. Three other British chemists, in particular, developed a keen interest. The first was T. M. Lowry, professor of physical chemistry at Cambridge. The second was Bernard Flurscheim, a somewhat enigmatic character. He had studied under Thiele—whom Robinson acknowledged as a source of his own inspiration—but had private means, and set up his own laboratory in Hampshire: he was thus out of the main academic stream. The third, Ingold, was the youngest of them all; he was appointed Lecturer in Chemistry in Imperial College in 1920, before moving on to be professor at Leeds in 1924 at the age of thirty-one. As we have noted earlier, he overlapped briefly later with Robinson at University College London.

Over a period of some three years (1923–6) these five engaged in heated controversy over their own variants of an electronic theory of organic reaction, conducted partly in meetings of The Chemical Society and partly in *Chemistry and Industry Review*, journal of the Society of Chemical Industry. By the beginning of 1926 some fifty communications had appeared in the *Review*, and the editor felt obliged to call a halt:

> Discussion of alternative polarities and kindred topics is of great importance, and we hope the letters we have published in these columns have cleared away a considerable mass of misapprehension. We are, however, obliged to wait for a period before dealing with the subject again: a proportion of our readers fails to understand the whole of the argument without a mental effort which is made unwillingly.[17]

The Chemical Society, too, was getting exasperated, and at about the same time Professor H. E. Armstrong (then retired from Imperial College) moved at a Council meeting—though he failed to find a seconder—that 'no further contributions to the mystics of Polarity will be received, considered, or printed by the Society.'

It would appear, however, that at times even the principals involved did not fully understand each other. At the height of the controversy Lowry wrote to Robinson, saying:

> I have not yet got used to your new system of arrows and want a little practice in order to think in terms of them. Even now I am not quite sure whether they represent a transfer of *one* electron or *two*. My rather vague impression is that in the earlier papers you generally transferred only one, while in your last C. and I. paper you appear to transfer two. It may not matter much which scheme is used but I want to get a clear picture in my own mind of your present view of this mechanism.
>
> I hope you don't mind being bothered like this with elementary questions; but in view of the fact that so many people misunderstand, more or less

willingly, the Manchester doctrine, it seems worth while to make sure that there are at least a few people about who willingly understand it.[18]

The last paragraph is an allusion to both Flurscheim and Ingold, but it was the latter who particularly incurred Robinson's ire. He did not become seriously involved until 1924, when he championed Flurscheim's theory of alternating affinities against Lapworth's notions of alternating polarities. In the summer of 1925 Robinson sent Ingold an advance copy of a paper—written jointly with J. Allan, A. E. Oxford, and J. C. Smith—which he had submitted to the *Journal of the Chemical Society*. It eventually appeared in 1926, under the title 'the Relative Directive Powers of Groups . . . in Aromatic Substitution'[19] as a commentary on three earlier papers. Ingold sent a friendly letter in reply:

I return your papers. Thank you for letting me see them. They represent in my opinion a very fine effort, especially on the theoretical side, and the theory is certainly one of *Organic Chemistry* and not of *Aromatic Substitution only*. Crystallisation in terms of electrons, as you have . . . [illegible] brought together. I find it easier to follow and I can now see in retrospect that its germination goes a long way back and is especially clear in your paper with Mrs Robinson of 1917 . . .
I shall also publish again on the subject but am not ready just yet. When I do I shall turn right round. I do not care two straws what the public in general . . .[20]

Unfortunately, Robinson at some stage mislaid what seems to have been a rather crucial page in the correspondence, but he clearly believed that Ingold now accepted his theoretical views in general, and acknowledged his priority, dating back at least to 1917, seven years before Ingold himself entered the arena. He was, therefore, incensed when Ingold's promised paper (written jointly with his wife, E. H. Ingold, like Gertrude a chemist) appeared,[21] claiming later:

When the promised paper by Ingold and Ingold appeared there was nothing in it which could justify a claim to novelty of outlook . . . Lapworth described reagents as anionoid or cationoid resembling those of chemically active anions or cations respectively. Ingold changed these terms to nucleophilic and electrophilic, but without the slightest alteration in the physical meaning. He adopted my method of representing a polarised system but as far as I know without adequate acknowledgement . . . he was apt to include a necessary reference to Lapworth or myself in a large number of references so that any idea that our contributions were original or specially applicable to the matter in hand was well and truly buried.[22]

In 1925 Ingold published the first[23] of a series of papers—ultimately exceeding thirty in total—with the general title 'The Nature of the Alternating Effect in Carbon Chains'. It was intended to provide experimental evidence for the validity of his own theory as opposed to that of Lapworth/Robinson.

Crucial, and also simple cases, in which the two hypotheses inevitably lead to opposite predictions, are not easily devised: but it is only by experiments on cases of this kind that an insight into the nature of the alternating effect can be gained.

It got off to a bad start. Ingold's prediction of the directive effect of a nitroso group was in accordance with the results of experiment; but in a joint reply Lapworth and Robinson demonstrated that in this instance their theory made the same prediction. It was a bad mistake, followed by another of a different nature later in the year, in Part III of the series. In this case (aminomethyl group) the error, as Robinson duly proved, was not in the interpretation but in the experimental work: Ingold was trying to move too fast, and was becoming careless.

However, Ingold was a good strategist—better than Robinson, as it turned out—and by the early summer of 1926 he had, without making it too obvious, distanced himself from Flurscheim and moved much closer to Robinson's nomenclature and symbolism. At a meeting of The Chemical Society in May of that year Robinson is reported as saying that:

The views now adopted by Ingold and his collaborators differed in no fundamental respects from those already advanced at various times by Lapworth and himself. The advocacy of non-polar theories of alternation, so characteristic of Parts I, II, and III of this [i.e. Ingold's] series had now been abandoned and this development was welcomed.

On the face of it, the enemy had surrendered—or at least agreed an armistice—but at this stage Robinson made a mistake. Satisfied with this seeming vindication of his views and claim to priority, he returned to his old love, research on the structure and synthesis of natural products: few of his later papers discussed the electronic theory of reaction as such. In this field Ingold seized the initiative, and in rapid succession published many more papers, in which he substantially followed Lapworth and Robinson, but put a gloss on their theory by developing a distinctive terminology of his own. Naturally, he came to be regarded as the front runner.

Belatedly, Robinson tried to retrieve the position by re-enunciating his views in lectures and review articles. Ill-advisedly, and surprisingly, he chose publications of rather limited circulation and influence. In 1932 he gave two lectures to the Institute of Chemistry, and these were published as a booklet.[2] Few people outside Britain, however, would have seen it. Significantly, it includes only one reference to Ingold's work—and that (without comment) to the aminomethyl group paper of 1926, in which his experimental work had been so grievously wrong. Today, this is a rare collector's item, and I treasure the copy Robinson gave me many years ago. Versions of this lecture were delivered in Belgium and Germany, and were duly published[2] in French and German, but again for a limited readership. Two years later he contributed a general account of his theory to the Jubilee issue of the *Journal of the Society of Dyers and Colourists*[24]; but although this journal was of high repute, its circulation was small and limited to specialists. For reasons not quite clear, a further paper prepared for the *Journal of The Chemical Society* early in 1938 never saw the light of day.[8] Ingold played his cards better. In 1932 he went to Stanford University on leave, and occupied part of his time in writing an extended review, 'Principles of an Electronic Theory of Organic Reactions', which was published[25] in *Chemical Reviews*, a prestigious journal read throughout the world, but especially in the USA. No doubt deliberately, remembering the published version of Robinson's 1932 lecture, this contained only a single reference to Robinson's work, though there were some to that of Ingold's old ally Flurscheim.

While Robinson and Ingold thus argued bitterly between themselves, both had to face a good deal of adverse criticism from the organic chemical world generally, which was sceptical of the new theories however they were presented, so much so that Ingold's 1934 exposition in *Chemical Reviews* was widely known abroad as 'The English heresy'.[26] By 1939, however, at the outbreak of war, the situation was very different: Ingold's ideas, and the terminology he had developed to express them, were generally accepted. There, for all practical purposes, the affair ended; but Robinson was reluctant to let the matter rest. In 1947 he once again set out his ideas in his Faraday Lecture to the Chemical Society;[3] but though basically a lucid exposition, its old-fashioned terminology must at points have baffled some of the younger members of his audience. Ingold really clinched the matter in 1953 with the publication in the USA of his monograph 'Structure and Mechanism in Organic Chemistry'.[27] This was widely read and accepted, and Robinson was often incensed to find that when he later advanced his own views in lectures in the USA and elsewhere members of the audience would tell him 'that the idea I had mentioned in the course of my talk had already been advanced in the monograph of Ingold.'[28]

The publication of Ingold's book spurred Robinson on to one last effort. He offered to me, as Editor of *Endeavour*, of whose Advisory Panel he had long been a member, an article on 'Some Intramolecular Electrical Effects on the Course of Chemical Change'. He deprecated this as being 'the mixture as before'; and this was indeed true. Obstinately, he clung to a terminology long since abandoned!

> Lapworth made the appropriate valuable extension of Berthollet's tables of affinity more than thirty years ago. He called the electron donors *anionoid* reagents and the electron acceptors *cationoid* reagents. By these terms he implied substances 'exhibiting a type of reactivity analogous to that of active anions and cations respectively'. He did not suggest that an anionoid agent must be negatively charged. Ingold later introduced the term 'nucleophilic' for anionoid and 'electrophilic' for cationoid. These are perfectly satisfactory equivalents, but the novelty is confined to the nomenclature, which expresses no fresh idea[4]

Having so uttered, he doggedly proceeded to use the terms anionoid and cationoid for the remainder of his article.

As was said earlier, the prolonged public controversy was not edifying. The essence of the matter was well expressed by John Shorter at the end of a well-balanced and scholarly review given in 1986 as part of a centenary tribute to Robinson organized by the Historical Group of the Royal Society of Chemistry.

> I must confess that I find the public discussion of the disagreements of these two great British chemists not long departed from us [Ingold died in 1970, five years before Robinson] somewhat distasteful and embarrassing. It has, however, been a necessary exercise to set Robinson's contribution to electronic theory in proper perspective. Such a discussion may also serve to remind us that chemistry does not just happen: it is made by real people, who in varying degree show not only admirable virtues but often show very human failings as well.[8]

8

Pre-war Oxford

Oxford was already familiar to Robinson, for over the years he had been a frequent visitor there. He had maintained close links with W. H. Perkin in the context of research, he had supervised research students, he had been an external examiner, he had attended lectures and conferences, and so on. Nevertheless, the peculiar constitution of the University, and its devotion to tradition, was to present him with professorial problems of a kind he had not previously encountered. As these peculiarities will be at least as unfamiliar to many readers as they were to Robinson, it is useful to digress briefly to review the Oxford scene in general, and the chemical scene in particular, in the early 1930s. This is very relevant, in that Robinson's activities were much influenced by the constraints within which he had to work.

While it is uncessary to consider the history of the University *ab initio*, it is pertinent to recall that the first firm historical date we have is in reference to University College in 1249, for this was the first of a series of colleges which are self-governing and yet are interrelated as corporate members of the University, which is in essence a federal union. The University is, indeed, something of an abstract concept, and most members regard their colleges as having the first call on their loyalty. To become a member of the University a person must first be accepted as a member of a College, and be presented by it for matriculation. In 1930 there were twenty-three men's Colleges, four women's Colleges, and one graduate College, All Souls: between them they had an undergraduate population of about 4000. To these must be added five small Permanent Private Halls. All graduates who hold the Oxford degree of Master of Arts belong to a body known as Convocation. This has in practice little power—save in electing a Chancellor—and the effective governing body is Congregation, which comprises those members of the academic community who are members of Convocation. In practice, administration is delegated to a small Hebdomadal Council.

Members of the academic staff at the level of lecturer or higher who are not Oxford graduates are admitted to membership of Congregation by granting them an M.A. by decree. Robinson was so admitted on taking up his appointment after paying the matriculation fee of £5! The studies of the University are organized in a number of Faculties—law, mathematics, physical sciences, etc.—and all members of the academic staff are normally members of at least one appropriate faculty. The faculty boards are co-ordinated by a General Board.[1]

Over the centuries study and teaching at Oxford were concerned largely with the learning of classical antiquity and the niceties of theology: as at other ancient universities, Latin was the common language of scholarship. The Church was a powerful influence, and the chapel was the centre of college life: college fellows were required to take holy orders, and remain celibate. As we have noted earlier, students were obliged to make a religious declaration as a condition of entry until 1871, within the lifetime of some of Robinson's new colleagues. And only ten years before that, at a meeting of the British Association in Oxford, there had been a famous clash between T. H. Huxley and Samuel Wilberforce, Bishop of Oxford, over the validity of Darwin's *Origin of Species*. Such an atmosphere of dogma was not conducive to the original thought that is the very essence of scientific enquiry. It was, of course, an attitude of mind not peculiar to Oxford, but one which permeated all the early European universities. In the seventeenth century Galileo was arraigned by the ecclesiastical authorities, and came within a whisker of losing his life, for supporting the heliocentric cosmology of Copernicus. It is, of course, idle to take the University to task for neglecting science during the first four centuries of its existence, for science in the modern sense is essentially a product of the seventeenth century: its full flowering can be identified with the founding of the Royal Society in 1660 and the Académie des Sciences in 1666. In this there is something of an irony, for the Royal Society derived from informal meetings of philosophers—many of them coming from London when this fell to the Parliamentarians—held in Wadham College under the auspices of the Warden, the mathematician John Wilkins. At this time, Robert Boyle formulated the famous law that bears his name. But with the Restoration the refugees returned to London, and the flame newly kindled in Oxford flickered.

In the eighteenth century chemistry was taught in a desultory way—largely by medical men, as was the practice elsewhere—and it was not until the middle of the nineteenth century that science was put on a proper footing. In the face of much diehard opposition Henry Adams—Reader in Anatomy, and later Regius Professor of Medicine—won approval for the building of the University Museum (so called because it housed many of the collections from the old Ashmolean) in the Gothic style, to house all the departments of medicine and science. It still dominates what is now known as the Science Area, (though the word Museum was used until 1964) on which are located all the principal laboratories. An enormous new reading room of the Radcliffe Science Library has lately been built under the front lawn.

At the same time, the University began to implement the recommendation of the Commission of 1854 to establish University professorships in science. Among these was the Waynflete Chair of Chemistry. It was named after William of Waynflete, a Lord Chancellor of England who founded Magdalen College in 1485: the professor was to be *ex officio* a Fellow of that College. The first Waynflete Professor was (Sir) Benjamin Brodie, an energetic man who was the first to occupy accommodation in the new building. He was followed in 1872 by William Odling, who was a gifted chemist, but during his forty years in office did little except deliver the statutory lectures required of him. During his time a new chemical laboratory was built in South Parks Road.

Odling was followed in 1912 by Robinson's mentor from Manchester, W. H. Perkin Jr. The appointment seems to have been made with minimum formality: the senior Oxford scientists consulted together, decided that Perkin was the man they wanted, and E. B. Poulter, Professor of Zoology, wrote to invite him to become the third Waynflete Professor. Perkin accepted, but made heavy demands. In particular, he wanted to build up a first-rate school of organic chemistry, and for this a new laboratory was necessary: organic chemistry then had only a few small rooms in the ground floor of the existing chemistry building. A promising start was made; Perkin was appointed in December 1912, and in February 1913 Convocation approved a grant of £15 000. Under Perkin's strict supervision plans were drawn up by Paul Waterhouse, with whom he had previously collaborated in the design of the Morley Laboratory in Manchester in 1909. Waterhouse was a highly regarded architect, and a Past President of the Royal Institute of British Architects. He used also the same builders—Armitage and Hodgson of Leeds—as had been employed for the Morley Laboratory. Not surprisingly the two buildings had much in common, including, unfortunately, extensive use of red brick and hard-wearing but unattractive brown and cream tiles. The style was decidedly Neolavatorial. The design was such that from his own room Perkin could look down on the teaching laboratories and see what was going on.[2]

There is, indeed, an interesting architectural genealogy here. The Dyson Perrins Laboratory embraced features of the 1909 Morley Laboratory in Manchester, which in turn was closely modelled on the 1870 Roscoe Laboratory there. In designing this last Roscoe had been much influenced by R. W. E. Bunsen's 1855 laboratory at Heidelberg, which was originally a monastery. Such a religious affiliation would not be out of place in Oxford. By a curious quirk—attributable to the romantic notions of John Ruskin, who had a considerable influence on the design—the part of the original Museum reserved for chemistry was a replica of the Abbot's kitchen at Glastonbury Abbey. Though quaint, this had some practical value: the high roof accommodated the fumes of chemical operations.

Perkin's new laboratory—which Robinson was to inherit—is today known as the Dyson Perrins Laboratory (always referred to in Oxford as the DP), and how this came about is an interesting tale.[3] In 1862 J. Dyson Perrins, a chemist, investigated the alkaloid berberine, extracted from the roots of *Berberis vulgaris* and other plants, and published its empirical formula in the *Journal of the Chemical Society*. Later Dyson Perrins became a partner in the firm of Lea and Perrins (his father), retail chemists in Malvern. Shortly afterwards they acquired from an ex-officer in the Indian Army—traditionally Marcus Lord Sandys—the recipe for a sauce; this became world famous as Lea & Perrins Worcester Sauce, and the partners became wealthy as a result. In due course Dyson Perrins' son, Charles William, came up to Oxford to read law at The Queen's College, and then, after a brief spell in the army, entered the family business. He had scholarly interests which his wealth allowed him to indulge. He was a shrewd collector of illuminated manuscripts, early woodcut books, and Worcester china. Many of his finest acquisitions are now in national museums and galleries.[4] It appears that Perkin knew the Dyson Perrins family quite well, and that the original link was the father's original paper on berberine, in which Perkin had been interested since his Heriot Watt days. In 1910, Perkin and Robinson—both then at Manchester—had published a joint paper (also in the *Journal of the Chemical Society*) establishing the structural formula for berberine.

The grant made by Convocation proved inadequate to complete the building, and in search of further funds it was natural that Perkin should turn to C. W. Dyson Perrins. He was well known as a generous philanthropist, his business had chemical connections, and he was interested in education: among other things, he was a life governor of Birmingham University. Perkin was not disappointed: Dyson Perrins made a donation of £5000.

Unfortunately, Perkin was even then not out of the wood. A prolonged building strike and the outbreak of war in 1914 caused delays, and costs rose. Although the building was finished, it was impossible to equip it adequately. Accordingly Dyson Perrins was invited to a small lunch party at Magdalen, and Perkin put his remaining problems to him. Again Dyson Perrins rose to the occasion: he not only gave £5000 for equipment but an additional £20 000 by way of endowment for future needs. Afterwards, he is reported to have said ruefully that it was the most expensive lunch he had ever eaten. Later, in 1919, the University recognized his generosity by giving him an honorary DCL degree. Perkin and his research workers eventually moved in at Easter 1916.

Some carvings in the fabric of the building are of interest. On the wall outside Perkin's laboratory was an inscription reading:

ALCHYMISTA SPEM ALIT AETERNAM

meaning Alchemy (chemistry) nourishes eternal hope. On the outside of the building is a small and fairly simple cryptogram linking the almost identically spelt names of Perkin and Perrins. Much more baffling, however, is one devised by Paul Waterhouse, the architect, who was a Balliol man. It reads:

BALLIOLENSIS
FECI
HYDATOECUS
O SI MELIUS

and close inspection reveals that certain letters (underlined above) are distinguished from the others by being longer. This proved too much for the classical scholar J. P. V. Dacre Balsdon, a Fellow of Exeter College, who concluded that it was 'the very acme of cryptographic lunacy'.[5] The true meaning was divined by one of Perkin's colleagues D. Ll. Hammick, a Fellow of Oriel, who was versed in Latin as well as in organic chemistry. It proved to be a punning reference to Waterhouse himself.

OF BALLIOL
I HAVE MADE (IT)
OF WATER HOUSE
O THAT IT WERE BETTER

The tall letters are Roman numerals, and if rearranged in descending order of magnitude read MDCCLLLLVVIIIII: that is, 1915—the year in which the building was completed. This amusing and harmless conceit is in the best Oxford tradition.

We must consider also the nature of the research school that he inherited; but first it is necessary to consider some of the other chemical laboratory facilities then available in Oxford. Originally, almost all the lecturing on science and laboratory work in Oxford was done in part of the Ashmolean Building in Broad Street (now the Museum of the History of Science) named after its principal benefactor Elias Ashmole (1617–1692). This continued in use until the opening of the Museum. But in the absence of adequate University facilities, a number of the colleges had made their own provision for laboratory work, and many of these were still in use in Robinson's day.[6] The Dr Lee's Laboratory in Christ Church, originally built for anatomy, was used for chemistry from 1767 to 1941. Charles Daubeny—distinguished for simultaneously holding three professorships, in chemistry, botany, and rural economy—became dissatisfied with his cramped accommodation in the Ashmolean Building, and at his own expense built himself a block of laboratories opposite Magdalen College: these continued in use for chemical teaching and research until 1923, but are now the College Bursary. Balliol and Trinity Colleges founded chemical laboratories in the second half of the nineteenth century, and these remained in use until 1942, when the new Physical Chemical Laboratory was completed. The Queen's College Laboratory was used for organic chemistry research from 1907 to 1934, when it was absorbed into the D.P. The last survivor of the college chemistry laboratories was the Sir Leoline Jenkins Laboratory in Jesus College: founded in 1907, it was not closed until 1946. These laboratories were anomalous, in that, although each was an integral part of a college, they were used by members of all colleges. I myself remember doing physical chemistry experiments in Balliol/Trinity, and going to lectures in Jesus.

During the war the original chemical laboratory in South Parks Road had been occupied by the Royal Air Force, and some of its inmates had found refuge in the D.P. When the RAF departed the building was refurbished and taken over by Frederick Soddy—famous in the context of isotopes—just appointed to the new Dr Lee's Professorship of Physical and Inorganic Chemistry. Thereafter it became known as the Old Chemistry Department (O.C.D.), prompting witticisms about 'Soddy and his old chemistry'. The refugees returned to their old home, and with the completion of the new wing in 1922 Perkin finally achieved his ambition of being head of the finest organic chemical laboratory in Britain. Robinson's inheritance was complete: it remains to be seen what he made of it.

Although Robinson himself had no direct connection with the college laboratories, they were an integral part of the chemical framework within which he had to work. It was an archaic system, which was workable only because it had evolved over a period of time, and those concerned were used to it. But it was not one which anybody would then have devised as appropriate to the needs of the day. To a newcomer it was bound to be perplexing.

Having thus briefly described the physical environment into which Robinson was to move, we must say something of those who worked in the D.P., and of how it was run. Although it was an active school of research in many branches of organic chemistry it was, of course, also an important teaching laboratory. These two facets were linked, in that the teaching staff (all then known as Demonstrators) all had their own postgraduate research groups, and most were Fellows of a college, acting as tutors to undergraduates there. In respect of their research, College Fellows were autonomous: the University required them to engage in research or advanced studies, and to co-operate with the Professor in running the laboratory. But neither they nor their research students had any obligation to participate in the Professor's own research. The undergraduate practical work was largely done in the two big teaching laboratories. Alongside the practical work there was, of course, a systematic annual programme of lectures.

In Perkin's day two important innovations were made. His view, shared by D. L. Chapman of Jesus College, was that no classified honours degree in chemistry should be awarded unless the candidate had carried out a research project, written a thesis, and been examined on it. From 1921 the course was thus divided into Part I and Part II, the latter being a fourth year devoted to research: this development ensured that Oxford graduates had a good practical as well as a theoretical background. A Part II thesis, if of sufficient merit, could qualify for a B.Sc. degree.

The second innovation was the D.Phil degree, introduced in 1920 as an alternative to the Ph.D which many chemists acquired at German universities before the war. In the immediate post-war years these were largely closed to students from Britain. This was a considerably more difficult proposition, being longer and involving more independent work with less supervision. Some candidates were graduates of other universities. These innovations gave a considerable boost to chemistry at Oxford, and graduates found themselves in increasing demand, especially for industrial firms such as ICI, founded in 1926.

Perkin got off to a difficult start, because the outbreak of war meant that, as in other universities, much academic research had to be set aside in favour of practical problems of national importance. He was, it will be recalled, in charge of the schemes by which 'colonists' were received—by Robinson among others while at Liverpool—from industry. The Oxford colonists included representatives from British Dyes, W. J. Bush, and Boake Roberts: there were also three refugee Belgian professors. This cross-fertilization had a useful long-term effect. The immediate post-war years were ones of strain for the teaching staff as men returned from the war; but by 1922, with the extension completed and the lodgers returned to the O.C.D., things were more or less normal. But by then Perkin was flagging. His age was beginning to tell—he was then 62—and this was reflected in the diminishing number of his research papers. As a consequence, fewer research students were attracted from afar, and the laboratory became somewhat inbred with Oxford graduates. In 1928 he became seriously ill, and there is some reason to suppose that he was the victim of mercury poisoning, due to his repeated recourse to what is known as the Emde degradation reaction, which involves use of sodium amalgam. He rallied, but died in Oxford on 17 September 1929 after an attack of pleurisy while on holiday in Switzerland.

In these latter years the informal manner in which he ran the laboratory made life easier for him. There was a minimum of paperwork, and the various research groups were left to carry on undisturbed unless they ran into difficulties. It is said that in this he modelled himself on the great Adolph von Baeyer, with whom he had worked in Munich in 1882–6: he regarded time given to anything except teaching and research as time wasted. Above all, he was fortunate in having Fred Hall as his laboratory steward and analyst. Fred ordered and dispensed chemicals and apparatus from his service room without so much as a chit—but he had a very shrewd idea about how much of everything was necessary, and was rarely deceived. He also did the book-keeping, and was responsible for the laboratory technicians. S. G. P. Plant and E. Hope—of whom we shall hear more later—did what was necessary in the way of academic administration. Hope took general charge of the laboratory during the one year interregnum between Perkin's death and Robinson's arrival.

To Oxford traditionalists the idea of declining an invitation to a chair there is almost unthinkable, and there is no reason to suppose that Robinson took long to make up his mind. Nevertheless, there is evidence from various sources that on two main grounds he was apprehensive about the change. The first was the college system, which is why it was discussed at some length earlier. A consequence of this was that the academic staff tended to follow their own lines, with their own students, rather than that of the professor. It was said, and rumours would certainly have reached Robinson, that in his later years Perkin found himself increasingly isolated within his own department. The second ground for doubt was that, again for the reason stated earlier, the D.P. housed a high proportion of Oxford graduates, who tended to be self-centred and unwelcoming to 'foreigners', of whom Robinson was to bring some twenty from London, and could be expected to import more over the years. We know from J. C. Smith[7] that in the months while he was waiting to take up his appointment he lost no opportunity to inquire into the working of the system. The importance it assumed in his mind is further indicated by the space he proposed to give to it in the second volume of his *Memoirs*: although this was never published, substantial parts exist in draft. Against such doubts, there were many advantages to be set. First and foremost—for a man devoted to research—were the facilities offered by a modern, well-equipped laboratory where he could at least expect a welcoming committee of four old Mancunian friends. Additionally, the prestige of the University, whatever its defects, was still very great, and Oxford was a particularly agreeable city in which to live.

In the event, a revealing remark shows how the pros outweighed the cons. When he arrived in Oxford he was forty-four, and in eighteen years had already occupied five chairs. He could, therefore, have been regarded as something of a nomad, and after a time somebody asked him when he would next be moving. To this he replied that, since the only better place to be was Heaven, he intended to stay in Oxford:[8] he stayed twenty-five years.

The four former Manchester colleagues we have already mentioned. They were Wilson Baker, J. M. Gulland (also with him at St Andrews), E. Hope, and J. C. Smith. Hope had worked with Robinson on alkaloids, and had come to Oxford in 1919: like Robinson, he was a Fellow of Magdalen. Unfortunately, his health began to fail in the mid-1920s, and he could give little time to research, and so could give Robinson little help in this respect, though he continued to lecture. J. C. Smith, a young New Zealander, had joined the D.P. from Manchester in 1928.

The remaining senior staff who awaited him, no doubt with apprehension matching his own, fell into two categories. One was what might be called the 'straight' organic chemists, interested, like Robinson, in experimental synthesis and degradation. The others were physical chemists interested in organic compounds. The leading figure in the first category was S. G. P. Plant, an Oxford graduate who had a rare gift for lucid lectures on stereochemistry. The physical organic chemists had particular cause for unease. Perkin had tolerated them, though he did not share their interests: how would the new man regard them? In the event, they continued more or less undisturbed. The 'sub-department' comprised N. V. Sidgwick, (one of the Waynflete Electors), D. Ll. Hammick, L. E. Sutton, and T. W. J. Taylor. Without exception they were Oxford men. Sidgwick had even been born in Oxford, and went to school there. He was a colourful character gifted with a mordant wit, not always appreciated by the victims of it, and an incisive writer. His memorial is *The Organic Chemistry of Nitrogen* (1910), subsequently revised by Baker and Taylor in 1937. Hammick, too, was colourful and versatile; he had come to the D.P. in 1921 after some years as a chemistry master at Gresham's School and Winchester. He was a good teacher, a great and convivial raconteur—a gift which found particular vent in describing his fishing exploits—and had a gift for composing limericks seemingly without effort. Taylor disliked demonstrating, but nevertheless did so very capably. He left Oxford when war broke out in 1939, and had a distinguished career outside the world of chemistry, as the first Principal of the new University College of the West Indies (1946), and then as Vice-Chancellor of Exeter University.

Nearly sixty years have passed since Robinson arrived in Oxford, but there are still survivors who can recall the occasion. One of them is Dr Muriel Tomlinson, who gave a vivid account in a lecture given as part of a Centenary Tribute to him which was embodied in the Annual Congress of the Royal Society of Chemistry held at Warwick in 1986.[9] She had then completed Part I of the organic chemistry course, involving nearly ninety preparations, ending up with indigo. Robinson took up office during the Long Vacation, and when she returned to start her Part II research year she found the laboratory transformed. In the large upstairs laboratory there were a score of strange research workers, a number of whom had come down from London with Robinson. A number were foreigners, including four rather forlorn Indians. Despite this outward and visible sign of change, in fact things went on very much as before, and initially Robinson was little in evidence. He did not take over the general course lectures in organic chemistry, but gave his own lectures, twice a week for three terms. In these a good deal of time was devoted to the Electronic Theory, and the remainder to heterocyclic chemistry—that is to say, to compounds having rings in their molecules incorporating sulphur, nitrogen, or oxygen: those discussed were both synthetic and natural. Like others who had been exposed to the same treatment elsewhere:

> I found the whole series fascinating, and the asides and extras which he interposed were often more rewarding than the main theme.

A year later, having finished her Part II, she moved up to the big laboratory to work with Robinson himself, and then it was a different story:

> . . . the last two years, working with Robinson himself, was happy and satisfying. We were a cosmopolitan crowd, as many of the pupils came from

overseas: I remember that at one time there were eleven different nationalities among the twenty of us in the big laboratory. The professor came to see us most days, and threw out suggestions so rapidly that I sometimes had to count them on my fingers to help me to reconstruct the programme. His insight into one's work was astonishing, but he could be disconcerting, and I remember once feeling very foolish when he replied to one of my suggestions 'I don't know that reaction'. If he didn't know it then it was certainly not a reaction. He hardly ever looked at the person to whom he was talking, and this caused great difficulty to one deaf lip-reader.

The reliance on foreign research students was, of course, an established practice. One of the most able, recruited by Perkin in 1928, was a Pole, O. Achmatowicz, who did outstanding work on strychnine. A glance at the names of co-authors of papers published in the early 1930s amply demonstrates Robinson's facility in getting on with foreign students: Sugasawa, Muraskami, Chakravarti, Fouseka, Seshadri, Suginome, Erdtmann, and Miki among others. The research was varied: while natural products—alkaloids, anthocyanin pigments, fatty acids, steroids, synthetic oestrogens (this last in collaboration with E. L. Dodds of the Courtauld Institute of Biochemistry)—held pride of place, there was also research on antimalarials (on which he had worked with Pyman at Boots). At Oxford, biochemistry was not established as a full Honour School until 1949: unlike many of his chemical colleagues, Robinson did not oppose this overdue move.

As we get an authentic glimpse into the laboratory, so at this stage it is possible to have a closer glimpse than hitherto of the Robinsons' domestic life, particularly through the recollections of his daughter Marion. At first they took a pleasant double-fronted house in North Oxford, 13 Norham Gardens, within easy reach of the D.P. There was an intitial contretemps as their cook, who had been with them in Manchester and London, at first refused to live in a house with this unlucky number. However, she had not to endure it for very long, as in 1932 they moved to a much larger, detached house at 117 Banbury Road. The Banbury Road house, much enlarged, is now the North Oxford Overseas Centre, a hostel for foreign students.

A large house was necessary to meet all the family requirements. Apart from the cook and other domestic staff, a nannie was necessary for Michael, Marion, then ten years old, attended Headington School. Gertrude regularly put in a full working day at the D.P., where she had been allotted two benches in what had been Perkin's laboratory. She started at nine o'clock and, apart from an hour's break for lunch, usually continued until seven o'clock in the evening.

Over the years Robert had acquired a massive collection of books and chemical journals—sufficient to make him largely independent of the well stocked Radcliffe Science Library—and one large room facing the Banbury Road was set aside for these and for writing. It was no unusual thing for passers by to see the light still burning there at two o'clock in the morning. In addition to all this, space had to be found for a variety of leisure activites. He enjoyed listening to music, and indulged himself in the best sound reproduction system available: his taste was catholic, ranging from Bach to Gilbert and Sullivan—even hymn tunes, without the words—and he built up a huge collection of records. As a performer, he was not as good as his brother Victor, but he enjoyed playing the piano for his own amusement. The large house also allowed him to indulge his photographic interest. The attic was divided partly into a darkroom, where he could do his own developing and printing, and partly into a playroom. In this context his chemical skill could be useful. The manufacturer of his favourite fine-grain developer refused (not unreasonably) to divulge its formula—so he simply had it analysed in the laboratory, and then made his own, at considerable saving in cost. Gertrude's main hobby was to renovate old furniture bought in the secondhand shops of Oxford and neighbouring Cotswold towns. There were, of course, limits to what could be acquired, and from time to time—occasionally to Robert's annoyance—familiar pieces were disposed of to make room for new treasures.

With both parents living such busy lives, it is surprising how much time could be found for family activities. On Wednesday evenings he might help with Latin homework. Saturdays often meant a visit to the Playhouse theatre, and on Sunday mornings Robert and Marion—in pre-by-pass days—would often cycle along the quiet country lanes around Oxford, or down to Wytham Woods and Godstow. Sunday lunch was always traditional, with a roast joint. Robert was fairly abstemious, but enjoyed a glass of hock with it. He was also a considerable smoker in his younger days, getting through a hundred cigarettes a day. The resulting smokers' cough led him to give it up for eleven years, but he fell again as a consequence of joining the ICI Research Council, to whom Sir Alfred Mond dispensed wonderful brown-green cigars. Rashly, he accepted the offer one day, and this was his downfall, for it led him back to cigarettes—though on a more modest scale. Occasionally there would be family excursions in search of wild flowers—an interest dating back to his student days at Manchester—or blackberries. Every summer there would be a rather grand garden party with strawberries, and occasionally smaller gatherings to entertain distinguished visitors.

Additional insight into family life in Oxford is provided by Robert's nephew, George Walsh, who came up to Oxford to read chemistry in the mid-1930s, until called up in 1940; he eventually got his degree *in absentia* as a prisoner of war in Germany. It was, he believes, Robinson's influence that secured him a place at St Peter's: Gertrude paid his graduation fees. Looking back, he recalls:

My uncle and aunt were always very kind to me and many of my school Holidays were spent at 13 Norham Road and later at 117 Banbury Road. Even earlier when I was 9 . . . I spent a month on the Isle of Arran with them (1927) and this was perhaps the finest holiday I have ever experienced. We had a rented cottage at Corrie . . . Marion and I spent a lot of time around a derelict boat near Brodwick. While I was at Oxford I spent most of my Sundays at No. 117 . . . Uncle R. was very much the family man, playing tabletennis or tennis with us. As he got older he developed a cunning slice on serving and always stayed on the winning side. He excelled in everything he did and was always the victor. He could lick me at chess with five or six pieces against my whole set, and I was too slow to remember his moves (I found his lectures equally difficult to follow and his publications, handed to me by my tutor, Mr Hammick equally hard to memorise).[10]

His niece Elizabeth Shirley Robinson, Victor's eldest daughter (now Mrs T. S. Sampson) also read chemistry at Oxford (St Hugh's College, 1943–52) and has similar recollections[11]

. . . the series of lectures he gave to undergraduates in their first year were the best attended of any that I went to—only to be compared with those of Professor Hinshelwood on Physical Chemistry. The lecture theatre was packed and some of the actual lectures were a *tour de force*—particularly those on the anthocyanins.

Apart from chemistry, his house at Banbury Road was always warmly welcoming—not only to me but to others as well, and my aunt was the greatest fun and a most stimulating person.

I particular remember my uncle for his wonderful expertise at chess and for his uncanny skill at solving mental puzzles—which he shared with my father.

He was always ready to communicate his expertise, whether in gardening—

chess—mountaineering—he taught my sister to climb—and he was *never* dull to be with.

Christmas remained a special occasion, with a great family gathering at Fieldhouse in Chesterfield for those who could manage it. (Robert had been appointed a Director of the family business in 1930, but was never active in its affairs.) There would be a huge Christmas tree, carol singers, party games, and all the traditional activities of the season. There were other visits to Chesterfield, and when these were made by car the journey was enlivened by singing songs from the Scottish Students' Song Book (perhaps a souvenir from St Andrews): in the intervals Robert, with his remarkable memory, would give long recitations from the Ingoldsby Legends, or Leland's Breitmann Ballads, burlesquing the German-Americans. On railway journeys, or while waiting in hotels or restaurants where such noisy enjoyment was inappropriate, no time was allowed to be wasted: home-made crossword puzzles were compiled and solved. Crosswords were indeed a perennial source of enjoyment, and that in *The Times* was usually at hand at mealtimes. Bridge problems had a special fascination for him, and he solved them seemingly without effort.

Robert was conscientious parent and—as increasingly happened as he grew in status—was a diligent letter writer when on his travels. He wrote particularly to his mother and sisters, to his brother Victor, and to his family in Oxford. Marion recalls, however, that, although he went to a great deal of trouble, his letters were sometimes disappointing: they tended to deal largely with mountaineering exploits or new chemical discoveries—neither of much concern to a young girl who had no great interest in either mountaineering or chemistry.

Indeed, Marion actively disliked mountaineering, and has vivid memories of bitter days in snow or rain, when her young arms and legs were too short to reach the foot- and handholds of the climb, and she might be left dangling in terror over a precipice. She states quite frankly that, understandably, some of her experiences gave her nightmares. Robert took the simple view that she had to enjoy it, as he did. However, these climbing expeditions were not wholly unenjoyable. There were cosy evenings relaxing by a log fire doing word games or identifying plant specimens collected during the day. Other days were spent in the mountains—usually in Wales or the Alps—simply walking, enjoying the scenery, and botanizing.

Robert was mindful of his own and other children in the family in other ways. At Christmas he would go to great trouble to find what he believed to be suitable presents for each individual. Surprisingly, in view of the zest with which he pursued walking and climbing on vacations, Robinson took little or no exercise in Oxford. Unlike Perkin, who regularly walked to the D.P. from his house in Charlbury Road, he always drove to the laboratory, though he always walked briskly up the stairs, scorning to use the lift. His choice of car was typical of the man. In the 1930s he had a very large Studebaker (a model prophetically named 'The President'), which he had to lay up early in the war as it consumed so much petrol—this excited a good deal of comment. Asked why he had such a huge car, he replied that it was entirely because of the size of its luggage compartment—it was the only car he could find that would accommodate the climbing gear of three people!

In Oxford he did, however, enjoy a fair amount of exercise and fresh air in cultivating his large garden. He had a special interest in shrubs and alpines, and devoted much time to building a rockery to provide a natural habitat for the latter.

Gertrude had her own car—a modest Standard 12, in marked contrast to Robert's Studebaker. She was a notoriously bad driver: her sight was poor, and she had a habit of slumping down in her seat and peering through the steering-wheel. Her short-sightedness began to be something of a problem in the laboratory, where she had some difficulty in reading thermometers and identifying small weights. Both of them were careless of their cars, and did as little as posible by way of maintenance, often neglecting even to keep the tyre propertly inflated. Surprisingly for the time, Marion was given her own car at the age of seventeen to take her to Headington School, her parents believing this to be safer than a bicycle. This relaxed attitude extended also to the laboratory. While the intensity of research, measured in terms of the number of first-class papers published in the chemical journals, was enormous, and the teaching excellent, Robinson had little interest in management, and neglected it accordingly. He was fortunate, therefore, in having in Fred Hall an experienced Steward devoted to the laboratory: he was correspondingly unfortunate in quarrelling with him at an early stage.

Hall had joined the Chemistry Department as a laboratory assistant as long ago as 1896, at the age of fourteen. Perkin realized his competence, and offered him the position of Steward in the D.P. when it opened. He had no formal training in chemistry; but he was a careful observer and quick to learn. He was familiar with all the routine experiments for the students' practical course, and when things went wrong he could often be as helpful as an official Demonstrator. Apart from this he could work independently, and he carried out the routine analyses that are an essential part of an organic chemical research. His wage was shamefully small, but he was allowed to charge five shillings for each analysis, thereby earning perhaps another £100 a year. With this background, it came as a surprise to many that when he died in 1962 he left a fortune of £36 000. The explanation was that over the years he had shrewdly invested a small legacy left to him by his father before the First World War.

J. C. Smith recognized his worth, and paid a warm tribute to him.

> . . . he reached the laboratory by 7.50. By noon he had usually carried out 2 or 3 analyses . . . he had supervised his assistants and the cleaners, ordered chemicals and apparatus, kept meticulous financial records, discussed analytical results, chastened students who had brought him impure specimens or handed in explosive substances without warning. Perkin gave him a free hand to run the laboratory and to rebuke and fine people who caused fires, broke apparatus, left water-taps running, or used the lift without permission. He was quiet, efficient, firm and fair; he was liked by nearly everyone and hated by the careless and wasteful. . . . I do not recall any unpleasantness between Fred and any of our Demonstrators.[12]

Clearly, such a man was of inestimable value to a professor who felt no inclination towards administration. Perkin had recognized this, and gave him a free hand. Robinson did not, and ill-advisedly spoilt a good relationship. One particular episode, trivial in itself, seems to have precipitated matters. A foreign research student complained to Robinson that his research was held up during the vacation because the Service Room closed at 4 p.m., and the outer doors of the laboratory were locked. With the hasty judgment to which he was prone Robinson concluded that Hall was taking advantage of a new professor, and closing the laboratory to suit his own convenience. Hall protested in vain that he was merely continuing a vacation routine established in Perkin's time, which had caused no inconvenience. All research workers were entitled to a key to the outer doors, and few found difficulty in estimating their needs for the rest of the day by 4 p.m. A further ground for dispute was that Hall continued to feel that, as under Perkin, he had to keep an eye on expenditure, which was rising alarmingly as new research programmes got under way. This, too, was ill received. Neither man was of a forgiving nature, and the rift was permanent, and damaging to the smooth running of the laboratory. However, this was not obvious to the casual observer. Muriel Tomlinson records that she was not aware of it until she returned to the D.P. after the war.

Until the opening of the D.P. the main centre for teaching and research in organic chemistry had been The Queen's College laboratory, presided over since 1907 by F. D. Chattaway, a Fellow of Queen's. When he retired in 1934—having already transferred most of his research to the D.P.—and the laboratory was closed, his steward, A. S. ('Gertie') Miller, joined Hall, and good-naturedly presided over the Service Room. To help cope with the growing demand for analyses G. Weiler and F. B. Strauss were installed with a laboratory of their own: previouly, hundreds of analyses had been sent to Schoeller in Berlin.

Fred Hall was so much a part of the D.P.—which he loyally served for nearly fifty years—that we should follow his life to its end, which was tragic. His wife died after a long illness, and after that he lost heart and began to neglect himself. In 1955 a new professor, E. R. H. Jones, arrived from Manchester bringing his own steward, George Ryder, with him. Fortunately the two got on most amicably, and Hall retired on half pay, coming in regularly each morning to do analyses with his usual accuracy: these hours in the laboratory, to which he was so attached, were his life. He was much touched when, in 1957, the University recognized his devotion by awarding him an honorary MA degree. But gradually his own health failed, depression overcame him, and in January 1962 he took his own life in his lonely home. It was a sad end for a man who, in his own sphere, had done so much for the D.P.

For Fred Hall the D.P. was the alpha and omega of life, but for Robinson it was very different. Muriel Tomlinson recalls that he came to see his research students 'most days'; but by 1935, when she left to go to Girton College, Cambridge, things were changing. He still did a little experimental work—though his genius really lay in generating from his immense store of knowledge and experience novel approaches for others to exploit under his guidance—but often his students, sometimes those most in need of help, saw nothing of him for a week or more, even when he was in the laboratory. J. C.

Smith opined that this was sometimes the result of a 'queer inhibition' which drove him into seclusion. However, over the period 1931–34 two of his immediate neighbours recall with pleasure almost daily contact. Most afternoons he would call on them at four o'clock, whereupon tea was brewed in beakers and they would make a combined assault on *The Times* crossword puzzle, interspersed with discussion of whatever chemical topic was then exercising his mind. One of the two was Alexander Todd (later Lord Todd), destined to go on to a career as brilliant as Robinson's. With J. W. (Kappa) Cornforth, who also worked in the D.P., but a little later, he wrote the official biography of Robinson for the Royal Society,[13] on which I have drawn substantially. The other neighbour was B. K. (Bertie) Blount, who also went on to a distinguished career in the army during the war, and in the government service.

There is some evidence, too, that by 1936, when Robinson reached the age of fifty, the chemical sparkle was diminishing. The instant diagnosis which could resolve a difficulty with which a student had been struggling for days or weeks still came, but a little less readily. This, too, may be a reason why he identified himself less with his research workers.

Although slightly out of chronological order, it is appropriate to mention here the comments of a distinguished Indian chemist, R. N. Chakravarti, who as a young man spent two years (1945–7) in the D.P. after gaining his D.Sc. at Calcutta University. His recollections of the D.P. laboratory in the immediate post-war years are of interest, the more so as he arrived as a stranger, with a very different background and no preconceived ideas.[14] The very circumstances of Chakravarti's arrival throw interesting light on Robinson's style. He had written also to Heilbron at Imperial College, one of his D.Sc. examiners, and had been told that all such requests to carry out research should be directed to the Office of the High Commissioner of India in London. Robinson, in contrast, promptly sent a brief but cordial cablegram: 'Very pleased to accept you as research collaborator after October this year.'

When the two first met they had a friendly discussion of his research programme, which—for reasons which will appear in a later chapter—was to deal with an important point of detail in the structure of strychnine. Thereafter it was a very different story.

Soon I came to know that he was a very moody and peculiar type of person and did not like any of the workers disturbing him in his room for any discussion about their difficulties. I never tried to verify this as I understood that he was keen for good results. At the beginning of the session in October he would go round each and every worker in the laboratories for a few days. During this period he would be at each work-bench to the full satisfaction of the worker. After this it was practically impossible to see him. Of course, I never experienced any such difficulty as I preferred to see him coming to the laboratory in search of me for the results. At times, he would suddenly come to one, have some talk regarding work which might be based on some of his rough (may even be termed as—crude) experiments and then swiftly leave the laboratory more or less looking at his footsteps, while a number of other students waiting all along the passage might fail to draw his attention. To be frank, at the beginning I was very sorry to see this state of affairs.

Grumbling was not infrequent. But one should have understood how busy he was with so many different types of engagements. Thursdays and Fridays he had to spend in London for meetings etc. Mondays were considerably devoted as a mail-day, with poor Mrs Geldart as his secretary, he preferred to write most of his ordinary letters in his hand writing. He was invariably present in the D.P. on Saturdays and often found working on Sundays. There was a large collection of students and well-experienced collaborators from different parts of the world and it was quite reasonable that the weaker students should mix with the senior workers for solving their difficulties instead of disturbing him.

In the face of such difficulty in discussing problems as they arose his research workers resorted to various ruses. One, much favoured by Chakravarti, was to leave on Robinson's table at night a written progress report and a statement of problems encountered. He would then deliberately absent himself—'mostly moving aimlessly around the outskirts of Oxford'—so that when Robinson, his interest aroused, came to seek him out he was not to be found. Thus frustrated, he would leave Chakravarti a written note of his opinions and suggestions. This elaborate manoeuvring had a threefold advantage. First, it avoided the immediate dogmatic response, which he would rarely retract, that was his instinctive, but not necessarily correct, reaction to a direct verbal question. Second, a written reply involved him in forming a reasoned judgment which he could not subsequently disown. Third, the device avoided the necessity of disturbing him, which he was known to dislike so much.

Yet once he was convinced that there really was a problem he would go to immense trouble to try to solve it—doubtless as much for his own satisfaction as that of his research worker. Chakravarti records:

Once he was convinced about the defect of his scheme, the worker would not only receive his due consideration immediately, in spite of his very busy time, but the status of the worker would also be considerably raised in his estimation. Robinson never had any dearth of time to meet such exigencies in spite of his very tight day to day programme. To him time was like a rubber balloon, which could be inflated as and when required to clear up any suddenly arising and unscheduled urgent work. Possibly that was the reason why he preferred writing his letters in handwriting to gain time—whereas for typed letters a dictation, waiting for the typed matter, correction of the typed matter, and then the signature.

Chakravarti left Oxford in 1947, to go to work with L. F. Fieser at Harvard; and there for the moment we will leave him. We shall encounter him again later, however, in a different connection, for despite the disparity of some thirty years in their ages this was the start of a lifelong friendship: in 1950 Chakravarti married Debi Mukherji, who had done a D.Phil. with Robinson in 1947–9. Over the years he received some two hundred letters from Robinson.

Robinson's failure to give sufficient attention to his department is apparent in another way. One of the responsibilities of the D.P. was to give service courses on organic chemistry to medical students, and this put a strain on the available accommodation, which was aggravated when the Queen's College laboratory was closed in 1934. Temporary relief was gained by putting up a single-storey wooden hut behind the D.P.; but this was no more than a palliative. In 1939, however, funds became available—in part from the Rockefeller Foundation and ICI—to build an extension, completed in 1940; and the history of this is curious. Robinson not only failed to consult with any of his colleagues—none of whom saw the plans until the first floor was actually under construction—but gave surprisingly little attention himself to its design. The evidence is that he wanted more space for his own research on biological aspects of organic chemistry, and simply told the architect to design an extension pretty much on the lines of the original D.P., conceived twenty-five years earlier. Architects not being familiar with the needs of contemporary chemical research, the result was, predictably, disastrous. A major defect was that the only link with the old building was a narrow corridor with eight steps in it, making it next to impossible to wheel heavy items of equipment, such as gas cylinders, from one to the other. There was no connection at all at first-floor level. All sorts of basic facilities—such as cold room and glass-blower's room—were lacking. Little attention was paid to fire hazards, always a major risk in this kind of research: the Robinsons had had a very serious fire in Sydney, and the D.P. itself had had many narrow escapes from disaster. The situation would have been difficult to retrieve at any time, and impossible after the outbreak of war.

In so far as comparable incidents are on record, we should not judge Robinson too harshly. When F. M. Brewer was designing the new Inorganic Chemistry Laboratory at Oxford after the war he was specifically told by C. N. Hinshelwood—the professor then responsible for both inorganic and physical chemistry—that he was to consult with nobody but himself. Not so far away, at Reading University, E. A. Guggenheim of his own volition went ahead and designed a new laboratory after the war with no reference to those who were to work in it.

Among the reasons for Robinson's detachment from laboratory affairs was certainly the growing intensity and diversity of his interests outside the laboratory. To some extent these were the natural consequence of his international reputation in the world of chemistry: both at home and abroad universities and learned societies were anxious to secure his participation in conferences and lectures. While at Manchester he had made several trips to India in response to invitations by his old friend J. L. Simonsen, who had played a large part in setting up the Indian Science Congress in 1914, and was its Honorary Secretary from then until 1926: in 1928 he was its President. He also had strong links with Switzerland, which accorded well with his mountaineering passion. Soon after his arrival in Oxford he was invited by Professor Paul Ruggli to spend a fortnight in the University of Basle, in November 1931, to give a course of six lectures: in return, Ruggli later came to Oxford to lecture on dyestuffs and intermediate products. The occasion proved an interesting one, quite outside the chemical context. Robinson stayed in the house of Dr and Mrs Hoffman-La Roche, of the famous Swiss chemical firm.[15] His hosts were devoted to modern painting, of which they had a considerable collection, while Ruggli was interested in music, especially that of Elgar.

Robinson had two other Swiss friends who divided their time between chemistry and art. One was Leopold Ruzicka, a Croat who was appointed professor in the Zurich Technische Hochschule in 1929: he was awarded a Nobel Prize in 1939. He formed a valuable collection of Old Masters, and often visited London to seek others among the arts dealers there. The other was Athur Stoll, Director of Research in Sandoz, another well-known Swiss chemical concern. He, too, formed a fine collection of pictures, which he housed in Lausanne. In such company Robinson's thoughts must surely have turned to his old colleague at Sydney, A. L. Lawson, who diligently sought Rembrandts and other Old Masters which he thought might have come to Australia with early Scottish settlers.

Over the years Robinson received honours in many countries; but the award of the Paracelsus Medal of the Swiss Chemical Society in 1939 must have given him particular pleasure. Long before this, however, he had been awarded high honours at home. The Royal Society, the oldest and most prestigious scientific society in the world, had elected him Bakerian Lecturer in 1929—when he chose strychnine and brucine as his theme—and subsequently awarded him its Davy (1930) and Royal (1932) Medals. The former, in bronze, is awarded annually for the most important discovery in chemistry made in Europe or Anglo-America: it was appropriate that Arthur Lapworth should win it in the following year. The Royal Medals are in gold, and two are awarded annually by the Sovereign, on the recommendation of the Council, for the most important contributions published in 'His Majesty's dominions' within the previous ten years: traditionally one is awarded in the field of the physical sciences and the other in the biological sciences. By a rare slip of the Society's pen Robinson's award in the next (4th, 1940) edition of the *Record of the Royal Society of London* was credited to Robert Robertson, who was in fact at that time the Government Chemist. The second Royal medal went to the discoverer of vitamin D, Edward Mellanby, then Professor of Pharmacology in Sheffield University, but shortly to be appointed to the highly influential position of Secretary of the Medical Research Council. During this period Robinson was also a member of the Council of the Royal Society, but not an officer.

While he was a committed academic, Robinson undoubtedly enjoyed the problems of the industrial world not only for their different intellectual challenge, but as a means of legitimately augmenting his income. Apart from his own comfort, he had the future of Michael to provide for. The demand for his services as a consultant was very welcome, and he showed considerable skill in avoiding a conflict of interests. Of his consultancies, the most important was that with ICI, which dated back to 1927, when Sir

Alfred Mond, first Chairman of the newly created ICI, set up a Research Council designed to link ICI's own research with that in the universities, thus beginning a tradition which has continued, in various forms, to this day.[16] Robinson was one of those who accepted the invitation to join the Council: others were F. G. Donnan (already a consultant to Brunner Mond, one of the original constituents of ICI) and F. A. Lindemann (later Lord Cherwell), who from 1932 was scientific adviser to Winston Churchill. ICI also fielded a powerful team, including three—F. A. Freeth, E. F. Armstrong, and Frederick Keeble—who were Fellows of the Royal Society. It was a powerful group, both in knowledge and influence, well qualified to promote

> the object with which Imperial Chemical Industries was organised . . . [which is] to place the Chemical Industry of the Empire in a position second to none in the world.

However, the scheme was not without its critics within ICI. Thus in 1936, Roland Slade, ICI Research Controller, wrote:

> When we look back upon what was decided as important fundamental work we find that it was always the work that they themselves were doing at the time.

However, the more influential G. P. Pollitt—Main Board director 1926–1945—saw this lack of obvious application as a merit of the scheme:

> Direct financial result from this type of research cannot be counted on: its chief objective is the training of research men and the investigation of problems which should eventually result in additions to our general chemical knowledge . . . The Research Council grant was not intended in any way to subsidise research having an immediate works value. It was ICI's main contribution to the raising of the standard of scientific work in the country.

Even if they had been aware of these cross-currents, the academic members had no need for concern: the scheme was secure as long as Mond was alive. Meanwhile they profited greatly from it. At the personal level, they were paid 50 guineas for each meeting—normally two a year—plus expenses: no doubt this was why the minutes show that they attended much more regularly than the ICI members! Additionally, their universities profited from regular grants for the purchase of chemicals and apparatus: in 1935 these exceeded £50 000, a not inconsiderable amount for that time. The main beneficiaries were the universities of Oxford, Liverpool, and Manchester, Imperial College, University College, London, and the University of Amsterdam. This last, seemingly anomalous, university needs a word of explanation. In the 1920s ICI had become interested in high-pressure reactions through working the German Haber-Bosch synthetic ammonia process. In this context they had, through Freeth, who had a strong interest in Dutch chemistry, acquired as a consultant A. Michels, of Amsterdam, a pioneer research worker in this field. Robinson was impressed by Michels, and in 1931 suggested that ICI should embark on a programme of high-pressure studies. While it is not to be supposed that this was accepted solely on Robinson's recommendation, it is a fact that in 1933 this programme adventitiously produced a novel polymer, polythene, which was to revolutionaize the plastics industry, and prove one of ICI's biggest money spinners.

Apart from this profitable association with ICI as a corporate body, Robinson was also consultant to one of its major Divisions, the Dyestuffs Division at Blackley, Manchester. There he attended monthly meetings of the Dyestuffs Group Research Committee (DGRC). This had wide-ranging interests, from dyestuff intermediates and the dyes themselves to chemicals of interest in the pharmaceutical and agricultural fields. The work for the DGRC continued until his retirement from Oxford in 1955.

He continued to be associated with Boots, and also became a member of a research committee set up about 1937 by the Anglo-Persian Oil Co. (later British Petroleum). This was in interesting contrast to the ICI Research Council. Originally, there had been a research committee headed by J. F. Thorpe, Professor of Organic Chemistry in Imperial College. However, Thorpe found himself in disagreement with A. E. Dunstan, the company's chief chemist, because the company would not support basic research, and resigned. The committee was reconstituted primarily to develop a process for making a high-octane motor fuel, by combining isobutane and isobutene in the presence of sulphuric acid. Apart from Robinson, the academic members of the new committee were Ian Heilbron (who succeeded Thorpe at Imperial College in 1938) and E. K. Rideal, Professor of Colloid Science at Cambridge. However, a company representative on the committee, F. Hilton,[17] recalls that within his experience (1937–9) the performance of both

Robinson and Heilbron at the afternoon meetings—held at the research station at Sunbury-on-Thames—was disappointing. They presided rather than initiated. One reason, he supposed, was that neither Robinson nor Heilbron were particularly strong on the relatively short straight-chain aliphatic molecules which are the main concern of the petroleum industry: their interests lay rather with the world of complex cyclic molecules. In Britain, as distinct from the USA, the organic chemical industry used coal-tar rather than petroleum as its principal raw material until after the Second World War. However this may be, it was the start of what was to prove a long and fruitful association with the petroleum industry, to which we will return later.

Additionally, in 1938 Robinson became consultant to Kemball, Bishop and Co., chemical manufacturers at Bromley-by-Bow in the East End of London. Their main products were citric and tartaric acids, which they made by fermentation. This connection was to prove significant in the development of penicillin during the Second World War, as Robinson put H. W. Florey in touch with the firm, which produced substantial quantities of crude penicillin, to be purified in the Sir William Dunn School of Pathology, Oxford. Robinson's role in the penicillin story we will come to later.

Preoccupation with College and University affairs was not among the major distractions from the D.P., for he took little interest in either, with the result that he made few real friends in Oxford. At Magdalen his new colleagues made him welcome. In the draft of the second (unpublished) volume of his memoirs[18] he recalls particularly the help he received from a disparate trio: the President, G. S. Gordon, formerly Professor of English Literature; G. R. Driver, Professor of Semitic Philology; and C. S. Sherrington, Waynflete Professor of Physiology. He was amused as well as interested by the proceedings of the Governing Body, where the business transacted was very varied because of the wide range of property owned by the college and its exercise of patronage in various directions. There were many trivia. He recalls that on one occasion some seemingly indefensible custom was to be discussed and, as he supposed, dealt with on the nod. Not at all: one Fellow, whose name he unfortunately could not remember, asked leave to speak:

> Mr President, I hope the College will proceed in this with the utmost caution, as this is the only first-class abuse now left to us.

The ritual of college dinners impressed him sufficiently to think it worth recording at some length. He describes how the Fellows in their gowns marched in procession to the High Table in order of seniority of appointment. Afterwards, gowns removed, they adjourned to the Senior Common Room for wine and dessert: he describes how in circulating the wine the gap in the circle caused by the fireplace was conveniently bridged by a little carriage activated by the weight of the bottle. At such gatherings, it was the duty of the two most junior Fellows—meaning the two most recently appointed—to offer round the wine and dessert. In his unpublished note he makes no comment on this, but in fact it is well known that he was not best pleased at the time, thinking such a task somewhat inappropriate to his status.

One College custom he viewed with mixed feelings. On Christmas Eve there was a special dinner, after which the choir sang the second part of the Messiah and—after a break—a series of carols. During the first part of these proceedings, lady guests were confined to the minstrels' gallery. During the interval, they were allowed to descend and partake of mulled port and sandwiches with the Fellows. Robinson disapproved of this segregation, and was pleased when the custom was changed, and women were allowed to take part in the whole of the festivities. He goes on to make a tongue-in-cheek comment on a further liberalization, which he had apparently learnt of by hearsay:

> Later there was another step, namely that Fellows may invite lady guests to college dinner at High Table, but I believe only on certain specified occasions, such as Sunday evenings [when rather few of the Fellows would attend anyway]. Of course we expect the ladies to be very flattered by this kind of attention, and it was thought that possibly there might be an inconvenient haste to take advantage of it. However it does not appear that either the hope or the fear were in any way justified.[19]

There were then, and to a lesser extent still are, fellows who dine in their colleges almost nightly; but this was never Robinson's custom. The College facility he most valued was the New Room, where Fellows could give luncheons or dinners to as many guests—including ladies—as the accommodation would permit. This privilege he made full use of during his tenure of the Waynflete Professorship, and occasionally afterwards, when he had moved from Oxford to Great Missenden and was privileged to be elected an Honorary

Fellow. Robinson was in no way a great college man, but after his retirement he acknowledged his attachment to it.

. . . after he had retired, he said to me how much he enjoyed his connection with the College and what an excellent institution a college is. We can hope and believe, therefore, that in this University and in this College that restless spirit found some part of the satisfaction that he was endlessly seeking.[20]

We need not doubt that he enjoyed College life, though perhaps not as fully as some—he once said he could argue with the Fellows of the Royal Society, but not with the Fellows of Magdalen; but it is doubtful if he got much satisfaction from the University. On the other hand a quotation from his unpublished memoirs—written in the mellower years of old age—makes it clear that he was by no means dissatisfied:

I never suffered any feeling that I was deprived of collaborators by the Oxford system. That was certainly due to the large number of brilliant men who came from other universities to work with me. An undesirable feature of the system was that it was hard on senior members of the staff who were not Fellows of colleges and who did not, at least in their earlier days, attract students from other universities.

He had, in effect, to tolerate what he could not change, as J. C. Smith relates.[21] Under the Oxford system the professor had no power to dismiss a University (as opposed to a Departmental) Demonstrator. On one occasion, at a meeting of the Board of Faculty, Robinson wanted to do just this, on the ground that the man in question had not done enough research. It was pointed out to him that under the terms of his appointment—'to engage in advanced study or research'—a University Demonstrator need do no research at all, he need merely study. There was a considerable storm, and Robinson proposed that the terms of appointment should be changed to make some research obligatory, but he got nowhere: the sheer inertia of the machine was too much for him.

One other Oxford institution which deserves mention at this point is the Alembic Club, which was formed in 1901 by a group of chemical undergraduates for informal discussion of a variety of chemical topics. Later, it became larger and more formal, and in Robinson's day the weekly meetings were held in the lecture-room in Jesus College. From time to time lecturers from outside Oxford were invited, and on some of these occasions Robinson would entertain them, and a dozen or more other guests, in the New Room at Magdalen: on other occasions, Sidgwick would do the honours at Lincoln. These were lively meetings, notable for the very active participation of undergraduate members of the audience. Afterwards, Sidgwick would invite the lecturer and a number of other people back to his room, in Lincoln, conveniently located just across the road, for drinks and more discussion. With typical Oxford male chauvinism, women were not admitted as members, though they might come to general meetings as guests of members: they were not even allowed to take University degrees until 1920. Not until 1950 was this anachronism removed.

The highlight of the Alembic year was the annual dinner, held in the Town Hall, one of the colleges or, later, as numbers grew, in the Randolph Hotel. In its heyday gatherings of 250 people were common. It was the occasion for a big reunion of Oxford chemists past and present, always with a strong contingent from industry, particularly ICI. The visitors spent some time in the laboratories, and this was a very valuable means of keeping Oxford chemists of all persuasions in touch with the outside world. The dinner itself was a splendid affair, and at the highest levels in the chemical world an invitation was an honour not to be declined. There were distinguished speakers, finishing with a final year undergraduate representing the Victims (of that year's examiners).

Such was the Alembic Club in Robinson's day; but, sadly, it has to be recorded that later the Club declined, and after 1966 the dinners—save for a single revival in 1973—were no more. The reason was that much of its success depended on the energy and organising ability of F. M. Brewer, the Reader in Inorganic Chemistry. To the shame, and disadvantage, of Oxford chemistry nobody was found to take his place after his sudden death in 1963.

9
The Second World War

By 1938 the inevitability of war with Germany had become generally recognized. Chamberlain's policy of appeasement, and his much quoted 'Peace in our time' message after meeting Hitler at Berchtesgaden and Godesberg in September carried little conviction in the face of continuing aggressive German policy. This included the earlier occupation of Austria in March, the occupation of the Sudetenland in October, and the invasion of Czechoslovakia in March 1939. In that same fatal month Hitler denounced Germany's non-aggression pact concluded with Poland in 1934; and within days France and Britain pledged themselves to support Poland. In August Germany signed a non-aggression pact with Russia; and on the 25th of that month a formal Anglo-Polish treaty of mutual assistance was signed in London. The mine had been primed, and it needed only the German invasion of Poland on 1 September to explode it. Britain and France declared war on Germany on 3 September, precipitating a Second World War barely twenty years after the end of the first.

Although a general policy of appeasement had been followed in Britain up to the Munich Conference of September 1938—when it was followed by an overt policy of rearmament—defence had not been wholly neglected, and various contingency plans had been made in the event of hostilities. It was widely recognized that a major factor would be heavy aerial bombardment of military, industrial, and civilian targets, as proved to be the case, though not as early as expected. As a defensive measure the principle of 'shadow' factories had been approved in 1934. These were government-owned factories for strategically important product, to be managed by industry on an agency basis in the event of war. For greater security they were mostly located outside the main industrial areas expected to be the prime targets. More important, however, was the development of radar to detect hostile aircraft at a distance and plot their progress. In Britain, this is particularly identified with Robert

Wason-Watt, who gave the first convincing demonstration of his system in the summer of 1935. Similar systems were being developed in other countries, but with less success: a vital feature of radar was that it could be operated under wartime conditions by non-specialists.

The man most particularly responsible for turning radar into a defence weapon that was to prove of crucial importance in the Battle of Britain in 1940 was Henry Tizard, an almost exact contemporary of Robinson (born 1885) who had read chemistry at Oxford as a demy (scholar) of Magdalen College. In 1929 he was appointed Rector of Imperial College, London, and in 1933 was invited to become Chairman of the Committee for the Scientific Study of Air Defence. In 1940 he led a Scientific Mission to the USA which gave a powerful stimulus to defence science there. Unfortunately, his political career then suffered an eclipse due to Churchill's close reliance for scientific advice on F. A. Lindemann (later Lord Cherwell), Professor of Experimental Philosophy at Oxford and Head of the Clarendon Laboratory. In 1942 Tizard came back to Oxford as President of Magdalen, in succession to Gordon, and thus—with their common interest in science—became a close associate of Robinson.

While Oxford University—even more so before the war than now—tends to be introspective and aloof from the affairs of the world, the gathering storm was apparent almost as soon as Hitler rose to power in 1934. A growing trickle of Jewish refugees was a reflection of the Nazi's particularly vicious persecution of intellectuals. Many of them were physicists, and Lindemann, in particular, seized the opportunity to strengthen his own department, which he had worked hard to develop over the years: when he took it over in 1919 the Clarendon Laboratory was almost moribund, virtually no research having been done there for half a century.

Indirectly, one of these refugee scientists was to have a considerable influence on Robinson's own research. In 1935 he engaged in a joint research programme, supported by the Medical Research Council, with Howard Florey, who had just been appointed Professor in the adjacent Sir William Dunn School of Pathology. Its purpose was to investigate the properties of an enzyme, known as lysozyme, which had the curious property of dissolving a number of species of bacteria. As a natural product, it appealed to Robinson. Florey had been interested in it for many years, but had hitherto lacked colleagues who could investigate the chemical nature of lysozyme. In his own laboratory he appointed as Departmental Demonstrator Ernst Chain, a refugee who had initially been accepted as a research student by Gowland Hopkins, the distinguished Cambridge biochemist. In the D.P. Robinson had a small team which included a D.Phil. student, E. P. Abraham. Good progress was made, and in 1937 lysozyme was obtained in crystalline form.[1] This was an achievement in itself; but its long-term consequences were more significant, for interest in lysozyme was one of the factors that led Florey and Chain in 1939 to embark on the historic research project—in which Robinson later became involved—that led to the development of penicillin as a uniquely powerful chemotherapeutic agent.

To this we shall come later, and meanwhile return to Robinson's personal situation as it was at the outbreak of war. He had then been in Oxford for nine years, and was firmly established both professionally, as Waynflete Professor, and domestically at his spacious and comfortable home on the Banbury Road. His daughter Marion—who had been sent to Headington School, Oxford—had gained a place at Lady Margaret Hall to read medicine, matriculating in 1939. After taking her first degree in 1942 she went to London to do her further training at University College Hospital, London, finally qualifying in 1946.

From the spring of 1939, however, Robinson had a major new preoccupation. Notwithstanding the minor irritation he had caused in the early 1920s through his controversy with Ingold, The Chemical Society had fully recognized his international reputation. In 1927 they had awarded him their Longstaff Medal, in immediate succession to F. G. Donnan, and in 1936 he was invited to deliver the prestigious Pedler Lecture, choosing as his topic Synthesis in Biochemistry. He had then already served one term (1921–3) as a member of Council and two (1926–9 and 1935–9) as Vice-President. It therefore came as no surprise to the chemical fraternity when he was elected President in April 1939 in succession to Donnan. The appointment was for two years, and was of particular significance in that 1941 marked the centenary of the Society, and major celebrations, with international participation, were in prospect. Sadly, these expectations were dashed by the war. In Robinson's second year of office no meetings, except those of Council and committees were held in London until—in very low key—the

Hundredth Anniversary Meeting was held there on 3 April 1941. In place of a glittering banquet a war-time lunch was served, with fewer than 100 Fellows, out of a total of nearly 4000, present. However, the one foreign guest able to attend was a man of great distinction: this was the distinguished American organic chemist, J. B. Conant, then President of Harvard University. Not until 1947 was it possible to organize a centenary celebration meeting appropriate for the occasion, by which time Robinson had moved on to even higher office, attending as President of the Royal Society.

Before considering the special circumstances of Robinson's Presidency it is necessary to define the role of The Chemical Society in the chemical world generally, and in that of Britain in particular. Founded in 1841, it was the country's premier society in this sphere, representing primarily the interests of academic chemists and others interested in pure research and teaching.[2] Its *Journal* had since 1849 been a leading forum for the publication of new experimental results—it was the one most favoured by Robinson—and it also published *British Chemical Abstracts* and *Annual Reports on the Progress of Chemistry*. Its offices in Burlington House, London, housed a large specialist library and a lecture room—it was in the latter that, years earlier, Robinson and Ingold had had so many of their controversial arguments over the electronic theory of reaction. The Chemical Society was not, however, the only body representative of British chemists. In 1881 the independent Society of Chemical Industry was founded, serving, as its name implies, the special interests of industrial chemists. Additionally there was the Institute of Chemistry, founded in 1877 to serve the professional interests of chemists in respect of such matters as qualifications for appointments such as public analysts, conditions of employment, remuneration, and so on. These three major bodies had fairly distinct spheres of influence; but there was, nevertheless, a certain amount of overlap: as a specific example we have seen how the Robinson/Ingold controversy spilled over from The Chemical Society into the pages of the *Journal of the Society of Chemical Industry*. Over the years there had, therefore, been some feeling that it would be logical to amalgamate them all into a single body representing all chemical activities. There were, however, practical difficulties about this, arising from differences in their charters, and—as is common in such circumstances—none of the bodies concerned were ready to face the forfeiture of independence that would be entailed.

Plate 1. Rufford House, near Chesterfield, birthplace of Sir Robert Robinson

Plate 2. A family group, *c.*1895: Robert with his parents and brother Victor

Plate 3. Robinson had a lifelong connection with the British Association, culminating in his Presidency in 1955. The announcement that Dublin would be the venue in 1957 prompted him to send this souvenir of the 1908 meeting to his friends as a Christmas card in 1956. With him is his friend John Simonsen.

Plate 4. The Chemical Collection room of Sydney University as it appeared in Robinson's time there (1912–15).

Plate 5. Gertrude Robinson in later life.

Plate 6. Tennis was a favourite pastime at the house in Banbury Road. Marion with the family dog, Pippa, in 1935.

Plate 7. Puzzles of all kinds were of perennial interest in the Robinson family. This crossword game was played by Robert and his brother Victor.

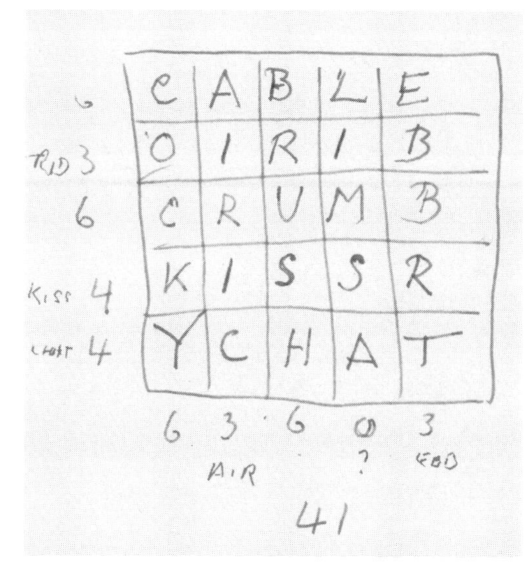

Plate 8. Robinson made no sartorial concessions to climate. This characteristic photograph—dark formal suit, trilby hat—was taken in 1959 in Israel on the occasion of the 25th anniversary of the foundation of the Daniel Sieff Institute.

Plate 9. In his Presidential address to the British Association in Bristol in 1955, Robinson made passing reference to the possibility of atomic explosives getting out of hand and causing catastrophic conflagration. This chance remark was widely seized upon by the popular press.

Plate 10. For many years, a regular feature of the British Association meetings was the presentation by the President of the Endeavour essay prizes for young scientists. Here Robinson shows the warm side of his nature in presenting one of the awards at the Bristol meeting in 1955.

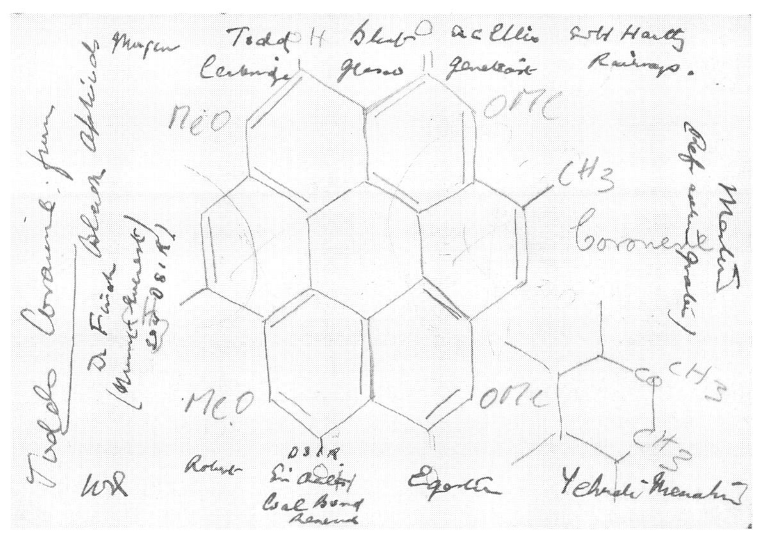

Plate 11. Robinson much enjoyed the meetings of the Royal Society Dining Club, which were cosmopolitan affairs. The back of this menu from May 1952 is characteristically covered with a scribbled chemical formula and the names of the other guests at his table.

Plate 12. Robert and Stearn enjoyed a long friendship with the Maxwells. This informal picture of the two Roberts and Stearn was taken on the Swiss side of the Matterhorn.

Plate 13. Robinson wholly relaxed among the craggy uplands he so much loved.

While the formation of a single chartered body lay in the distant future—achieved in fact with the foundation of the Royal Society of Chemistry in 1980—the economic problems of the early 1930s, and a decline of membership, did lead to some compromise moves. In particular, 1935 saw the formation of The Chemical Council, embodying representatives of each of the three bodies as well as three members of the trade organisation, the Association of British Chemical Manufacturers (ABCM).

The principal aim of The Chemical Council was to:

> establish a fund having for its sole object the allocation of grants for the co-ordination of scientific and educational publications, the publication of new discoveries, the promotion of research, the maintenance of a library, and the provision of adequate accommodation . . . The fund is intended to support the publications of the Chemical Society and the Society of Chemical Industry, and the library of the Chemical Society, the great British chemical library—one of the finest in the world—to which all chemists in this country have access.

The declaration of intent made it clear that the Council was looking for substantial industrial support, hence the involvement of the ABCM. It was not disappointed: its first appeal, launched in June 1936, raised over £50 000, a very considerable sum for those days, when the Chemical Society's total annual income was only £18 000. In addition, ICI had since 1932 made regular annual contributions of around £500 to the cost of Chemical Society publications. Thus Robinson was fortunate in inheriting a more favourable financial position than his immediate predecessors.

He was also the heir to an important organizational change adopted in 1938. Until that time the Society, although strictly national, had been London-oriented: the only geographical concession to Fellows was that those living more than fifty miles from Charing Cross paid a reduced subscription. Under the new regional system the British Isles were divided into six constituencies—chosen to coincide with county boundaries, to avoid confusion—each having a specified number of representatives on Council in proportion to the number of Fellows resident in it. The first Council elected on this territorial basis took office in March 1939, and thus Robinson was the first to preside over it.

Towards the end of Robinson's Presidency another important reform was introduced, known as the Joint Membership Subscription Arrangement. According to this, chemists could, on advantageous terms, extend their membership of one society to cover one or two of the others. The effect on the Chemical Society was little short of dramatic. In the first year of its operation 122 Fellows rejoined and 861 were elected, representing an increase of nearly a quarter. Income rose in proportion, from £20 000 in 1940 to £25 000 in 1945.

The full benefits of these changes were not realized until the Society could resume its normal activities after the war. Much of Robinson's term of office had to be devoted to giving effect to contingency plans. In common with other buildings in vulnerable areas those of the Society had to be protected as far as possible against damage by bomb-blast and fire. Crates of the most valuable books were dispatched for safe storage in Bangor, where Robinson's old friend J. L. Simonsen was Professor of Chemistry. N. V. Sidgwick, his colleague in the D.P., who had been President 1935–7, arranged for the accounts office to be moved to Lincoln College, Oxford, and to accommodate the whole of the headquarters there if necessary.

Meanwhile the society carried on its work as best it could. Very early in the war an Advisory Research Council was set up, with Robinson as chairman, to organize university research workers to involve themselves in chemical problems of national importance. By the end of Robinson's tenure of office nearly 200 such problems had been taken up, and liaisons had been established with the three principal national research organizations—the Department of Scientific and Industrial Research, the Medical Research Council, and the Agricultural Research Council. These activities were tiring and time-consuming, demanding frequent visits to London under very difficult circumstances and in blackout conditions at night.

Robinson cannot be said to have left his mark on the Chemical Society during his tenure of office, but whatever his intentions might have been the circumstances of war made anything more ambitious than survival out of the question. Nevertheless, he could at least point to a gratifying increase in membership, from 3695 in 1940 to 4516 in 1941, thanks in large measure to the joint subscription scheme. But the big changes—more library space and better staff accommodation—still lay far in the future. They could not be achieved until the Royal Society moved from Burlington House to sumptuous new accommodation at Carlton House Terrace in 1967: the Chemical Society was then able to expand into the vacant accommodation.

Robinson's commitment to the Chemical Society ended in April 1941; but although the relief was considerable, he still had many calls on his time in connection with wartime needs. One such was chemical warfare, which the Germans had introduced during the First World War in 1915 in the form of chlorine gas. This had limited success, however, partly because it was subject to the vagaries of the weather, and partly because effective gas-masks were quickly developed. In July 1917 the Germans introduced a new chemical weapon, in the Ypres sector. This was mustard gas, the vapour of which attacks the lungs, while the liquid produces severe and dangerous blistering. In Britain, this created something of a panic in Whitehall, and the Ministry of Munitions set up an *ad hoc* Advisory Panel of twelve scientists to recommend defensive and retaliatory measures. At that time, however, Churchill was planning a thorough revision of the Ministry, to which he had just been appointed; and as a firm supporter of chemical warfare he set up a formally constituted Department of Chemical Warfare. Within this new organization W. J. Pope, Professor of Chemistry at Cambridge, soon came to assume a leading role, and played an important part in developing a process for the manufacture of mustard gas in Britain.

Between October 1917 and February 1919 hundreds of different substances were investigated for possible use in chemical warfare. Interest then lapsed, especially in view of the Geneva Protocol of 1925 prohibiting the use of chemical warfare. This, however, was so loosely drawn—and automatically became invalid if any combatant failed to observe it—that all major governments felt obliged to maintain at least a watching belief. One development observed with keen interest was the development of organophosphorus insecticides by Bayer in Germany in the mid-1930s. A by-product of this was the discovery of Tabun, the first of the so-called nerve gases. Under the Nazis, large-scale production began in Silesia in 1942.[3]

In the event, however, chemical warfare was not used in the Second World War, but all concerned were obliged to take precautionary measures. Thus it was that Robinson came to be associated in the inter-war years with Pope as a member of the Chemical Board. He later followed Pope as chairman of two technical commitees, respectively responsible for chemical studies of new substances and for their manufacture. In the same context he became associated with Victor Rothschild, a distinguished zoologist of Trinity College, Cambridge: during the Second World War he had a distinguished career in military intelligence, and was a bomb-disposal specialist in the British Intelligence Service. Between them they were called on to advise the Home Office on whether gas-masks should be issued to the civilian population. Their considered opinion was that it was unnecessary—as the event proved—but they were over-ruled. This decision indirectly brought Oxford into the picture in a different way. Much of the practical work on the absorbent filling for civilian gas masks was done by Robinson's neighbour in the Physical Chemistry Laboratory—Cyril Hinshelwood—assisted by B. Lambert, a Fellow of Merton College.[4]

Apart from this advisory work on chemical defence he was involved also with other committees concerned with the manufacture of explosives (Ministry of Supply) and the supply of personnel for ministry staffs (Ministry of Labour). One such commitment involved him in a question being asked in Parliament. A committee chaired by Sir Edward Appleton, concerned with the allocation of scarce materials for the construction of chemical plant, included himself and Heilbron, then Professor of Organic Chemistry at Imperial College. They were asked to consider proposals put forward by Henry Dreyfus, chairman of British Celanese. It involved some novel processes which they regarded as impracticable 'paper syntheses', and they reported accordingly. To the annoyance of the committee Oliver Lyttleton, the Minister of Production, acceded to the request; but in response to a question in the House was obliged to admit that he had failed to take authoritative scientific advice.[5]

During the war private travel abroad was, of course, virtually impossible; and for such an inveterate traveller as Robinson this must have been a considerable disappointment. Nevertheless, in spite of the difficulties and dangers, travel on government business still went on. Thus at the end of 1941 he found himself dispatched to the West Indies, with his old friend and colleague John Simonsen, to explore the possibility of new science-based industries to create local employment. This had some important practical results. On their return, the Colonial Products Council was set up, with Simonsen as Director of Research and Robinson as one of the members. Later, the Tropical Products Institute was established in London. A major objective was to explore the possibility of using sugar—rich in carbon—as a raw material for the organic chemical industry. In support of this a research group was established in the Chemistry

Department of Birmingham University, under Dr (later Professor) Maurice Stacey. Robinson took a keen interest in this work, which was eventually transferred, under L. F. Wiggins, to a new Sugar Technology Research Institute in Trinidad.[6]

Simonsen died in 1957, and Robinson wrote his obituary for the Royal Society.[7] In this he recalls a similar trip with Simonsen in 1944, when they visited the USA and the Caribbean, partly in the company of Alexander King, then head of the UK Scientific Mission in Washington. The party had intended to go to Jamaica, but all available aeroplanes were grounded; so they went instead to British Guiana, and had long discussions with senior government officials and industrialists. They were impressed by the potential availability of water-power and the deposits of good-quality bauxite suitable for the manufacture of aluminium; the latter were subsequently developed by Canadian mining interests.

They were no less impressed by the high incidence of malaria in the fertile coastal strip, and immediately sent a cable to Heilbron requesting dispatch of DDT, the powerful insecticidal properties of which had been discovered in 1939 by P. H. Müller in Switzerland—research for which he was awarded a Nobel Prize in 1948. A vigorous campaign was duly launched—in spite of the objections of the local Italian malariologist, who had no faith in chemicals—with such success that 'The infantile mortality rate was reduced to such an extent that a new labour problem may menace the future.' Even the sceptical malariologist was converted, and sent an enthusiastic report. This interest in a local malarial problem was no doubt prompted by Robinson's long-standing interest in synthetic anti-malarial drugs. The visit to Washington had been largely to co-ordinate the work of American and British Committees organizing research on this problem.

They then went on to Puerto Rico, visiting the struggling University of San Juan, then being remodelled to include provision for research. When Robinson mildly suggested that it would be a good thing to invite American aid he was soundly reprimanded:

> . . . the reply was that this would be quite unnecessary and that the teachers at present in the University would be quite capable of initiating research. I do not know with what success this jump into the deep end was made.[8]

Moving on to Trinidad they relaxed 'playing a kind of golf-croquet on the Governor's extensive and rough lawn.' They had, however, more serious things to do. They visited the Imperial College of Agriculture, and found themselves in agreement with others that it was not fulfilling expectations. They recommended the foundation of a Microbiological Research Institute, which duly materialized under the auspices of the Chemical Laboratory—counterpart of the National Physical Laboratory—at Teddington. It did good work, though a project to produce a protein-rich yeast was not a success: Robinson records that 'this was found to be nauseous, so far as human consumption was concerned.' Their visit was also productive in that it led to the establishment of a Sugar Research Association to assist local planters and processors. They also visited the Great Bitumen Lakes at La Brea.

At first sight such a tour might seem not too arduous, especially as it was a relief from the discomforts of wartime Britain. In fact, those who have experienced such fact-finding advisory tours, being received and escorted by a succession of hosts anxious to see that every waking hour is filled, will know that they can be very tiring. By way of relaxation they were to go to St Thomas, in the Virgin Islands, where they were to be the guests of the Governor. The reason for this choice is not clear: Anthony Trollope, described it as 'one of the hottest and one of the most unhealthy spots among all these hot and unhealthy regions.' They had an initial disappointment: the Danish chauffeur who met them—the islands were once Danish territory—announced that they could not after all stay at the Residence. Instead, they were taken to a small hotel called Bluebeard's Castle. There they were received by the Chief Justice of the island, which doubtless appealed to Robinson's sense of dignity. The Chief Justice informed them that one of his privileges was to determine when Parliament should meet. Apparently he deemed one o'clock in the morning appropriate, an example of authoritarianism that clearly appealed to Robinson's sense of humour.

With his heavy load of committee work and time-consuming excursions such as these, it is not surprising that his colleagues and research students at the D.P. saw rather little of him. There teaching proceeded much as usual, as it was government policy to maintain as far possible the supply of trained scientists and engineers from the universities. In the winter terms, however, the practical programme had to be revised, as it was impracticable to black out the big teaching laboratory. Although some basic research continued, much effort had to be diverted to work of national importance, in particular that on synthetic antimalarial drugs already mentioned above. The capture of the East Indian plantations by the Japanese cut off supplies of the natural drug, quinine, and the two synthetic alternatives (mepacrine and pamaquin) were not wholly effective, and had toxic side-effects, including intense yellow staining of the skin in the case of mepacrine. The need for a better drug was urgent, for the Allies had to maintain large armies in some of the worst malarial regions in the world. In the event, the solution was found not in the universities but in the laboratories of ICI. They introduced a novel product, proguanil (paludrine), which proved highly effective in the later stages of the war. It was a complex substance, containing both sulphur and chlorine in its molecule.

Earlier than this, however, Robinson had become deeply involved in the chemical investigation of a chemotherapeutic agent of a far more powerful nature. This was penicillin, which had been discovered by Alexander Fleming at St Mary's Hospital, London, in 1928, but not pursued, because he failed to realize its true significance. In 1939 Howard Florey and E. B. Chain, in the Sir William Dunn School of Pathology, Oxford, quite independently initiated a general research project to investigate antagonisms between micro-organisms, and among the substances they selected was Fleming's penicillin. The story of how this led to the advent of penicillin as a uniquely valuable antibacterial drug has been the subject of much study, and therefore need not concern us here.[9] Instead we will concentrate on Robinson's own contribution.[10]

Briefly, the background is as follows. In 1938 E. P. Abraham—who had been working in the D.P. on lysozyme, and then moved on to a peptide-protein project—left Oxford to spend a year with Hans von Euler at his Biochemiske Institut in Stockholm. After war broke out in 1939 he made his way back with some difficulty to Oxford, but found that the protein work had been abandoned. As an alternative, Robinson suggested that he should move down the road to join Florey and Chain in a study of wound shock, a long-standing medical problem which had assumed new importance with the likelihood of many military and civilian casualties during the war. However, this project was short-lived, as the basic premise proved unsound; and so Abraham became involved, most immediately with Chain, in the penicillin project. By May 1940, thanks in large measure to an ingenious extraction process devised by N. G. Heatley, sufficient penicillin had been laboriously accumulated to make it possible to conduct some crucial experiments on mice infected with lethal doses of streptococci. These were dramatically successful; but, as Florey wryly remarked at the time, 'a man weighs 3000 times as much as a mouse', so that clinical trials were totally impossible with the quantity of material then available. Further animal experiments were equally convincing, and the experimental results were published in the *Lancet* of 24 August 1940. The next step was to produce enough penicillin to treat human patients—first achieved in Oxford in the first half of 1941, using small quantities of precious penicillin laboriously accumulated in the Sir William Dunn School of Pathology—and Florey had high hopes that the *Lancet* paper would arouse the interest of some of the chemical and drug companies capable of working on the necessary scale. In this he was disappointed. France had fallen, and Britain stood alone; the battle of Britain was won, but the Blitz began the day before the paper was published. The war at sea was intensified, and in 1940 no less than 3½ million tonnes of British and Allied shipping were lost. British industry was struggling to survive, and was in no fit state to embark on what was still—despite these early successes—a speculative enterprise involving unfamiliar technology. In the event, Florey had to seek the help of the chemical industry in America, making an extended tour there, with his colleague N. G. Heatley, in the summer of 1941. This had considerable initial success, which was much stimulated when the USA entered the war after Pearl Harbor in December. This led to a crash programme for penicillin production by several leading American companies and government agencies, which ensured that sufficient penicillin was available to treat all military casualties from D-Day onwards, then less than three years ahead.

All this penicillin was produced by fermentation, just as in Florey's laboratory in Oxford. The difference was that the industrial production was on a far greater scale, and used very sophisticated technology, resulting not only in vastly higher overall yield, but, by using selected strains of mould and modified culture medium, a very much greater (ultimately as much as 20 000 times greater) production per unit volume. All this involved a heavy investment in money, manpower, and equipment for which there were many competitive demands. It was undertaken because in the circumstances of war this was a race against time: only fermentation offered the certainty of providing the penicillin needed within a short space of time.

However, it was recognized that there was an in-built risk in this policy. Research chemists might not only unravel the structure of penicillin, but devise a means of synthesizing it economically, thus making the fermentation plant obsolete virtually overnight. This was more than a hypothetical possibility, for the history of the chemical industry records plenty of precedents. Thus the advent of synthetic indigo in 1897 virtually destroyed the indigo plantations in India and elsewhere in the Far East. In the special circumstances of war such an outcome would be desirable, in that it might make penicillin generally available more quickly and cheaply. There was, therefore, a powerful incentive to launch a massive programme to determine the chemical structure of penicillin, a necessary preliminary to its synthesis.

As we shall see, this programme became a huge international venture; but it was inevitable that its beginnings should have been in Oxford. For a time, this was the sole source of penicillin, and Florey had immediately to hand two highly skilled chemists—Chain and Abraham. These he had recruited—Chain several years earlier—in the firm belief that medical research problems would yield only to interdisciplinary attack. They were not, as is often supposed, recruited as part of some kind of penicillin taskforce. Additionally Florey had Robinson as his immediate neighbour, a scientist recognized not only as one of the greatest organic chemists of his day but one with a particular interest in natural products. They had already collaborated over lysozyme, and although Robinson was held in some awe by his contemporaries, Florey was in awe of nobody: it was a logical and fruitful association. Among Robinson's colleagues was, in particular, Wilson Baker—who had worked with him in Manchester—who had a particular interest in the synthesis of natural products, and was skilled in the manipulation of molecules with unusual structures.

Although more extensive clinical trials had to await the availability of relatively substantial quantities of penicillin, much less was required for chemical investigations, and an early start was possible. Even so, strict economy was necessary, and experiments were commonly conducted with samples weighing only a few milligrams. Heatley was particularly resourceful in devising techniques for handling such small quantities. Small, it should be stressed, is to be taken in the context of the time: today results are obtained with far smaller samples, sometimes literally only a few molecules. For

Robinson, association with this microwork was a new experience: he was notoriously given to working on a grand scale.

Initially, results were confusing, because the material was impure. It was impossible to distinguish between the reactions of penicillin and those of associated impurities, which in even the best early material were later found to amount to 93–99 per cent. By the middle of September 1941, however, when Florey returned from his 'hard-selling' tour of American industry, material of about 50 per cent purity was available. With this some more meaningful work could be done, and it was at this stage that Chain and Abraham established the formal collaboration with Robinson and Baker. A few months later the group was joined by J. C. ('Kappa') Cornforth, an Australian chemist who had lately gained a D.Phil. degree under Robinson. In addition, Dorothy Hodgkin, of the Department of Crystallography, provided much help through use of the then relatively new technique of X-ray crystallography. At that time she needed only a minute amount of a crystalline material quickly to determine the molecular weight of a degradation product and whether or not it was identical with a synthetic product. In 1945, she and her colleagues went on to confirm unequivocally the structure of penicillin deduced from chemical evidence. For this—a formidable task in pre-computer days—and other pioneer work she was awarded a Nobel Prize in 1964. The citation stated:

> You have solved a large number of structural problems, the majority of great importance in biochemistry and medicine, but there are two landmarks which stand out. The first is the determination of the structure of penicillin, which has been described as a magnificent start to a new era of crystallography. The second, the determination of the structure of vitamin B_{12}, has been considered the crowning triumph of X-ray crystallographic analysis.

Very soon, however, the chemical investigations extended far beyond the confines of Oxford, eventually involving 130 British scientists in 11 research groups and 300 American scientists in 22 research groups. As the full implications of penicillin became apparent there were fears that it might be developed in Germany and the other enemy countries, so strict security measures were enforced on both sides of the Atlantic from the early part of 1943. British reports were submitted first to a Penicillin Production Committee of the Ministry of Supply, and later to a Committee on Penicillin

Synthesis set up under the Medical Research Council. There was a corresponding system for collating American reports. Circulation of these reports—some 750 in all—was, of course, strictly limited, and very few people saw them in their entirety. The full scope of the chemical work was not generally apparent until 1949, when the full chemical story was told in a massive work of over 1000 pages: very appropriately, Robinson was one of the three joint editors.[11]

The coordination of this big Anglo-American programme required much administrative organization, and this was much facilitated during the course of a visit made to America by Robinson in 1943, in the company of Heilbron, who had a penicillin research group working at Imperial College, London. In the present context, however, we must limit ourselves largely to the fortunes of the small Oxford group. This, in fact, got off to a rather ragged start, due to what seems, in retrospect, to have been an elementary analytical blunder. When Robinson first took an interest in penicillin in 1943 one of his first questions was whether penicillin contained sulphur. He was told that, although this had been identified in early samples, it had been reported absent in later ones, known to be purer because of their higher antibacterial activity: penicillin appeared, therefore, to contain only carbon, hydrogen, oxygen, and nitrogen. In the summer of 1943, however, Wilson Baker and Dorothy Hodgkin found it impossible to reconcile this with some newly reported analytical results from a characteristic degradation product of penicillin isolated in crystalline form by Abraham. This revived the possibility of sulphur being present, and it is interesting that—after all the sophistication—Baker demonstrated this by a simple test that could have been carried out in the nineteenth century. He fused a little of a penicillin derivative with sodium, and placed the product on a silver coin: a tell-tale black stain of silver sulphide gave irrefutable evidence of the presence of sulphur. Conventional analysis then confirmed the pressure of a single sulphur atom in the molecule. Wilson Baker recalls that when he telephoned Robinson to tell him this momentous news he replied magnanimously: 'This is a feather in your cap, Baker.' This cleared up various misunderstandings, and progress continued. The identification of sulphur was of particular significance to Dorothy Hodgkin, for this is a relatively heavy element, producing a distinctive focal point in the X-ray crystallographic pattern. A month later came another surprise. Purification of penicillin had so much improved that chemists in the laboratories of the Squibb Corporation—one of the leading American manufacturers—had succeeded in producing crystals of penicillin sodium salt, which hitherto had been known only as a gummy, resinous substance. In accordance with agreed procedure this news was at once passed on to Oxford through the American Embassy in London. Abraham immediately converted a purified preparation of penicillin to sodium salt, and duly obtained crystals which he took to Dorothy Hodgkin for X-ray analysis. This showed beyond doubt that American penicillin and Oxford penicillin were not the same! The difference was small—a minor variation in a side-chain due to the use of different culture media for growing the mould—and did not affect the main results. It was significant, however, in showing that high antibacterial activity was not identified with a totally specific chemical structure. After the war, this observation was exploited in the development of a range of semi-synthetic penicillins, in which the changes were rung on the basic molecule by deliberately altering the side-chains. This led to products with significantly different clinical properties.

At this point some personal differences of opinion emerged, refuting the general supposition that science is cold and objective and not prone to subjective influence. The details need not bother us unduly, but the situation was such that the structure of penicillin had been almost, but not quite, unravelled. On the basis of the available evidence Robinson put forward what is called the thiazolidine-oxazolone structure, but Abraham considered that this was not wholly consistent with all the experimental evidence. As an alternative, he proposed what is known technically as a β-lactam structure. This was at once supported by Chain and Baker. In October 1943 Robinson prepared a report for the Penicillin Production Committee, and left it for the other three to approve and dispatch while he was away from Oxford for a few days. In his absence, they had the temerity to add the β-lactam formula as an alternative before sending it off. Predictably—for we have already noted his antipathy to having his chemical judgements questioned once he had uttered them—Robinson was extremely angry, and sent in a personal disclaimer stating that he believed the β-lactam structure 'somewhat improbable.'[12] In Oxford, passion ran high, especially in the case of Chain, who—in a very different way—could be quite as temperamental and controversial as Robinson. It was widely reported in Oxford that one encounter between the two ended with Robinson throwing an ink bottle after Chain and shouting after him: 'I do not want to see that wretched little man again.' However, passions cooled, and in the course of time the rift was healed. Elsewhere, reaction was more dispassionate; but most chemists involved in the project favoured the Robinson oxazolone structure, though a few short-lived alternatives to both this and the β-lactam structure were advanced. Extensive research along conventional lines failed fully to resolve the controversy, but, as we have noted, in 1945 Dorothy Hodgkin, and her colleague Barbara Rogers-Low, unequivocally demonstrated by X-ray analysis that Abraham's β-lactam structure was indeed correct.

Despite so much effort, the main practical objective of this exercise was not realized. Although parts of the penicillin molecule were synthesized, no acceptable total synthesis was achieved before the Anglo-American enterprise was wound up in 1946. Not until some ten years later was success achieved, by John Sheehan of the Massachusetts Institute of Technology. But his synthesis was only of academic interest, and industrially uneconomic: to this day, all penicillin is manufactured by fermentation. In 1979 the Royal Society organized a two-day symposium entitled 'Penicillin Fifty Years After Fleming.' In the course of this Sheehan remarked:

> It is remarkable that research in the penicillin area has continued for 50 years and has produced such spectacular benefits to mankind . . . The penicillin molecule remains unique in intensity of research lavished upon it and in the substantial claim that penicillins are indeed the queen of the antibiotics and continue after 50 years to occupy a pre-eminent position in medicine.

Penicillin was certainly by far the most important and unusually constituted substance with which Robinson was associated in a long life devoted to the chemistry of natural products.

In Europe, the war ended with the surrender of Germany on 8 May 1945. In the Far East, however, it still continued with no clear end in sight. In the event, however, it was quite shortly to be ended—dramatically and, later, controversially—with the dropping of atomic bombs on Hiroshima and Nagasaki in the first week of August. The war years had been ones of strenuous activity for Robinson, and despite obvious tasks ahead—such as getting his laboratory back on a peacetime footing, and absorbing an inevitable influx of ex-servicemen—he might reasonably have looked forward to a fair measure of relaxation. In reality, he was on the verge of a new and demanding phase in his career, which will be the main topic of the next chapter.

10

The last years of academe

As every soldier carries a Field Marshal's baton in his knapsack, so ambitious professionals of every kind look to achieving some ultimate pinnacle which demonstably stamps them with the approbation of their peers. In politics it is to be Prime Minister; in the Church, Archbishop of Canterbury; in the judiciary, Lord Chancellor; and so on. In British science, it is to be President of the Royal Society. It was to this high office that Robinson was elected in 1945; but to understand how this came about, and its significance, something must be said about the nature of the Society and its organization.

The Society was founded in 1660 by a small group of natural philosophers who had earlier been meeting informally in Oxford and London. It received a Royal Charter in 1662, and is the oldest scientific society in the world: it is also acknowledged to be the most illustrious. Membership is select and correspondingly coveted: most of the Fellows are from Britain or the Commonwealth, but a few Foreign Members are elected each year, 'from among persons of the greatest eminence for their scientific discoveries and attainments'. Traditionally, the ruling monarch is Patron of the Society, and there is special provision for the election of Princess of the Blood Royal, Foreign Sovereign Princes, Peers of the Realm, and Privy Councillors. In short, it is a very august body, and takes itself very seriously.[1]

Over the centuries the Society had had its ups and downs, especially in the late eighteenth century and the early nineteenth, when it came to be dominated by aristocratic dilettantes. Matters came to a head in 1830, when the astronomer J. F. W. Herschel was passed over as President in favour of the royal Duke of Sussex. This led to a radical revision of the statutes in 1847, introducing strict regulations for the admission of Fellows, which had become almost entirely casual. Slowly, the Society regained its former lustre as a body representative of all aspects of British science.

By the twentieth century the Society was recognized as the premier national academy of science, and as such was formally consulted by the government on all political matters with scientific connotations, and was responsible for the allocation of certain funds for research and other purposes given as a Parliamentary grant-in-aid. Despite this official recognition, it remained a strictly autonomous, self-perpetuating body. At the end of the Second World War the Fellowship numbered 450, and 25 new Fellows were elected annually. It elected its own members, from which were appointed a governing Council consisting of 21 Fellows. This in turn elected a President, a Treasurer, two Secretaries, and a Foreign Secretary.

The office of President was an onerous one. On the one hand he was responsible for internal affairs—scientific publications, administration of research funds and the government grant-in-aid, presiding at meetings, entertaining distinguished foreign visitors, and so on. Then there was responsibility for the accommodation in Burlington House, Piccadilly, which the Society had occupied since 1873, and for a considerable administrative staff. Ex officio, the President had to represent the Society at many major scientific gatherings both at home and overseas, and to respond to many invitations which, though essentially personal, were nevertheless derived from the office he held. In all this the President had, of course, the support of his officers and staff; but nevertheless there was a heavy load of personal responsibility. Additionally, of course, he had normally to discharge simultaneously the obligations of a senior academic appointment. Nor was the burden short-lived. Although technically elected for one year, it had long been tacitly assumed that the President would in fact hold office for five years.

Among all these various obligations, one is of overriding importance—the Anniversary Meeting held on St Andrew's Day, 30th November. In the afternoon there is the Annual General Meeting, at which the President distributes various medals to their winners, each with a laudatory citation. This is followed by the Presidential Address on some scientific theme. The big event, however, is in the evening, when there is a banquet of several hundred people in one of the great West End hotels—usually the Grosvenor House. Apart from Fellows and their ladies there is an impressive list of guests. This normally includes the Prime Minister of the day, several senior ministers, members of the diplomatic corps, senior industrialists, and so on. It can fairly be called a glittering occasion. Until fairly recent times full evening dress with decorations was *de rigueur*, though today dinner jackets are permissible.

The President's after-dinner speech is a matter of much consequence, for this is a unique occasion for bringing to the notice of a very influential audience matters of major significance to the scientific establishment. What he says is widely reported, as is any ministerial reply. Only on one other annual occasion does such an opportunity arise. This is the summer soirée in June—again a full-dress occasion, with ladies present—at which guests view a number of scientific exhibits and regale themselves with champagne and a cold buffet. This is an excellent occasion for a little quiet lobbying of a less formal nature.

Having said this, it is apparent that the President must be a man of parts. He must not only be a scientist of world stature, but a man able to deal with government at ministerial level and with the captains of industry. He must also have the ability to express his views incisively, but yet avoid giving offence to those with whom he has to associate. Not surprisingly, Presidents have not always possessed all these desirable qualities in full measure, but the Society went to considerable trouble every five years to elect one who at least seemed to have the necessary potential.

Leaving these general considerations aside, what was the particular situation in 1945 as Sir Henry Dale neared the end of his term of office? He had successfully steered the Society through the difficult years of war, but his successor had the equally onerous task of leading it through the no less difficult years of post-war reconstruction. Ultimately, such a choice must be subjective, representing the individual view of the electors. Nevertheless, certain criteria could be invoked as guides, and it is instructive to look at the records of Robinson's five predecessors—Dale, Bragg, Gowland Hopkins, Rutherford, and Sherrington. Their average age at election was 64—though this includes Bragg, unusually elderly at 73, and Rutherford, a youthful 54—and on average it might thus be expected that their successor would be around 60. In this context modern circumlocutions about gender can be forgotten: the next President would certainly be a man, for women were not admitted until 1945, the last year of Dale's term of office. Dale and three of his four predecessors had been Nobel Laureates when elected: the exception was Sherrington (1920–25) who gained a Nobel Prize in 1932. Thus a Nobel Prize was not a *sine qua non*, but a candidate who lacked it would certainly have to be judged as potentially in that august league. The electors would also naturally take into account previous experience as president of a major learned society. Some civil distinction, such as a knighthood, would also be a valuable asset. Finally, of course, the electors would closely scrutinize the candidate's record within their own Society, looking for such pointers as the award of one or more of their prestigious medals—especially the Copley Medal—or delivery of one of the important endowed lectures.

Over and above all these considerations was a crucial determinant intrinsic to the Society's organization. For many years it had adopted a dual approach to science: on the one hand there were the physical and mathematical sciences (A), and on the other the biological sciences (B). The firmly established convention was that the Presidency alternated between the A and B disciplines. As Dale was basically a physiologist it followed that his successor must be drawn from the ranks of the physical scientists. No biologist, however strong his claims, could be a contestant. He would have to wait another five years for his claims to be considered—and then he might fail on the ground of age, or for other reasons.

Although for months previously there is much speculation in scientific circles about the outcome, the election process is as secret as that for a Pope. It is, nevertheless, interesting to consider some of the possible candidates whose chances were being rated in the Common Rooms of Oxbridge and elsewhere. Looking back over a period of almost half a century, where the principals concerned are long since gone, it is difficult to assess the weight that might be given to personal qualities, but it is clear that there were several whose claims were comparable with Robinson's.

One who stands out in particular is W. N. Haworth, whom we have already encountered as a fellow student of Robinson's at Manchester, and formerly a member of the department of chemistry at St Andrews. In 1920 he had been appointed Professor of Chemistry at Birmingham, and in 1945 was Director of the large and flourishing department of chemistry there, with an international reputation in the carbohydrate field. He had been a Fellow since 1928, and the quality of his work had been recognized with the award of the Davy Medal (1934) and a Royal Medal (1942). Only the previous year he had been invited to deliver the Bakerian Lecture. In 1937 he had shared the Nobel Prize for Chemistry with the Swiss chemist Paul Karrer. He was a man of great energy, with considerable experience of government and other committees. Haworth himself cannot have failed to recognize that all this made him a strong candidate; but yet he had deliberately made himself unavailable, by accepting in 1944 the Presidency of the Chemical Society for a two-year stint—which in itself indicates that he was not averse to holding high office of this kind. It may well have been that he simply preferred the bird in hand. No doubt he was conscious that with the Society's strong Oxbridge/London sympathies his Birmingham base would count against him, as might his notorious feud with James Irvine of St Andrews, another Fellow and Davy Medallist. In his *Memoirs* Robinson recalls an attempt by the Oxford chemists to effect a reconciliation:

> The bad relations existing between Irvine and W. N. Haworth became notorious, and I rember very well a dinner at the Athenaeum club, which was organised by N. V. Sidgwick, and other well wishers of both, with the idea of bringing this distressing state of affairs to an end. After dinner Irvine spoke first. He adopted a most conciliatory tone and acknowledged in gracious terms the very great contribution to sugar chemistry by W. N. Haworth. Haworth replied with his thumbs in his vest pockets and spoke of a clash of temperaments. He did not really budge an inch. Afterwards, E. F. Armstrong said that he supposed that we had temporarily buried the hatchet, but the haft was showing above ground and could be grasped at any convenient moment.[2]

However, as this feud was no more notorious than that between Robinson and Ingold, it could scarcely, in the event, have been of critical importance.

Another name which must have come up for consideration was that of Sir Henry Tizard, whom we have also already met as Robinson's new President at Magdalen College, Oxford, and previously (1929–42) Rector of the Imperial College of Science and Technology. He thus had high standing in the academic world, and was one of the rare scientists who had long and close experience of what C. P. Snow has called 'The Corridors of Power'. In the post-war world this would have been a great asset. At sixty, he was of an appropriate age; but since his election to the Society in 1926 they had given him no recognition—no office, no medal, no lecture. We must suppose that in his pre-war work on radar—which had been crucial to Britain's survival in the Battle of Britain—he had strayed too far from the groves of academe to be acceptable to the Society, which collectively had little sympathy with applied science. This despite the words of its original Royal Charter that its 'studies are to be applied to further promoting by the authority of experiments the sciences of natural things and useful arts'.

Comparable claims could be advanced on behalf of F. A. Lindemann, Professor of Experimental Philosophy at Oxford, and a Fellow since 1920. He was then fifty-nine, and in 1940 he had become Churchill's personal adviser on scientific matters, effectively displacing Tizard, and had served in the coalition government as Paymaster-General (1942–5). He was clearly a man with great influence, though less so since the fall of Churchill in the 1945 election.

Another possibility mut have been Sir George Thomson, son of the famous 'J. J.' of Cambridge (who had been President 1915–1920), a Fellow since 1930, a Nobel Laureate (1937), awarded the Society's Hughes (1939) and Royal (1949) Medals. During the war he had been closely associated with the development of atomic energy. At fifty-three he might perhaps be considered a little young for the Presidency, but only a year younger than Rutherford was when he was elected in 1935.

How the matter was finally debated we cannot know, but Robinson himself throws some light on it in a note prepared for the second (unpublished) part of his *Memoirs*. At that time he was serving his second term as a member of Council, and was thus party to the preliminary discussions concerning a successor to Sir Henry Dale:

> The question of his successor was carefully studied at a number of meetings and in the last of these I found I had been nominated myself and was asked to leave the room while the discussion proceeded. In accompanying me to the secretary's office, Sir Henry said 'I don't think they will appoint you' but offered no further explanation. After what seemed a long time I was called back into the council chamber and was told that the Council would be pleased if I could find it possible to accept the nomination of Presidency . . .
>
> It has certainly not been a matter over which I have lost sleep but I have wondered from time to time how this election came to pass. There were several scientific cliques supporting leaders in war service but unfortunately their loyalty was unidirectional. A process of cancellation allowed me to slip in on grounds of achievement in scientific researches. I had already been awarded the Davy and Royal Medals and in 1942 the Copley Medal. The latter is the highest medal awarded by the Society in scientific research and has come to be regarded as a qualification for nomination to the Presidency.[3]

Clearly, Robinson was genuinely surprised that he should have been elected, and said as much in a letter he promptly wrote to his mother.

The St Andrew's Day banquet at which he was soon called upon to preside was the first since 1939, and the principal guest and speaker was the newly elected Prime Minister, Clement Attlee. In his reply, Robinson acknowledged his own unpreparedness:

> My situation tonight resembles nothing so much as that of a child thrown into the water in order to teach him to swim. I do not mention this by way of apology, but rather of explanation. At this stage, I am hardly well qualified to speak with authority of the plans that have been forming in the minds of those whose duty and privilege it has been to look ahead for the well-being of the Society

Nevertheless, he did not hesitate to raise immediately a question which was to preoccupy him for the next five years, namely the need for better and more spacious accommodation for the Society:

> The Royal Society has grown out of its clothes and needs a new suit. We are well aware that, for reasons known to all of us, the necessary coupons will not be available for some time but the case is desperate . . . Let us choose our material, get measured, and hope for a good place in the waiting list . . . I cherish the ambition that during my term of office the father of scientific academies, this Royal Society of London, will be able to perform its functions with the maximum of convenience, and will be housed in such a way as to symbolize the unchallenged prestige which it enjoys throughout the civilized world.

Sadly, his ambition was not to be realized: not until 1967 did the Society move from Burlington House to its present sumptuous accommodation at 6 Carlton House Terrace. It was a move which caused much controversy within the Society, and was one about which Robinson had great misgivings, for his own plans—and those of Dale, his predecessor—had been very different.

Their idea, which had a substantial measure of support, was that a Science Centre should be established on the South Bank of the Thames opposite Somerset House. This would house not only the Royal Society but all the other principal scientific societies, the research councils, and perhaps the Patent Office and the Department of Scientific and Industrial Research. It would also be a conference centre able to accommodate large international gatherings. At first, the scheme gained an encouraging measure of support in both the scientific community and the government: both Stafford Cripps, President of the Board of Trade, and Herbert Morrison, Lord President of the Council, were in favour. A Representative Scientific Committee was set up so that the views all interested parties could be voiced. But there was also opposition: many Fellows were reluctant to share facilities with what they regarded as the *hoi polloi* of British science, and—on the grounds of personal convenience—were reluctant to move far from the amenities of the Athenaeum Club. They preferred to stay at Burlington House, where the discomfort was felt largely by the staff, rather than move to the distant *terra incognita* across the river. By 1950 Robinson had to admit that his hopes had been 'pathetically optimistic': nevertheless he continued his campaign until 1953, during the Presidency of Edgar (later Lord) Adrian.

On this particular aspect of Society policy Adrian found himself in some embarrassment. While he was not in favour of a Science Centre on the South Bank, he naturally did not wish to seem at odds with his immediate predecessor. By way of compromise, he drafted a carefully worded memorandum which he sent to Morrison. Parts are worth quoting as revealing not only the grounds for opposition to Robinson's scheme, but the complacent self-esteem of the Society:

> Its present rooms in Burlington House in the heart of London are not magnificent, but at least they emphasise its independence, one might almost say its aristocratic seclusion . . . It would soon forfeit its reputation if it lost any of its independence and its aloofness from sectional interest, if it did not occupy a position of dignity in a building worth of its unique status in the scientific world.
>
> The ideal solution would be a completely separate building . . . Its premises should be self-contained: though they might share services like heating with the rest of the buildings they ought not to share corridors, staircases and entrance halls . . . The Society must not be expected to share its lecture hall with other societies except as a favour . . . In Burlington House the Royal Society has no difficulty in maintaining the independence and almost Olympian dignity and exclusiveness needed for a supreme scientific council . . . It ought not to accept anything less than the best that the site can offer for the whole scheme will suffer if the Royal Society cannot play its part effectively in premises worthy of its pre-eminent role.[4]

Small wonder that when Howard Florey—a robust outspoken Australian—became President in 1960 he is reputed to have said that 'he hoped to get something done even if he had to carry the Royal Society kicking and screaming into the twentieth century'!

After this, the scheme languished, especially as the government announced in December 1952 that the planning of the Science Centre had had to be deferred in view of the need for economy. Nevertheless, there remained in existence a Scientific Societies Accommodation Committee (SSAC), of which Robinson was a member. When Adrian informed the members of the shelving of the scheme, he got a very cool reply from Robinson:

> . . . The outcome is very disappointing since it amounts to a complete shelving of the scheme; not even planning is in progress. The need for economy is obvious but the application of the principle is arbitrary and relative. Many of us think that the encouragement of science should have a higher priority than is granted . . . Opponents of the scheme are now able to say 'I told you so' and exert an influence altogether disproportionate to their numbers. I do hope you can do something to neutralise the depressing activities of the retrogressive element among us.

There for all practical purposes, the scheme foundered: even the SSAC was disbanded. It briefly flickered into life again as part of a proposed redevelopment of Covent Garden, but that was short-lived. In 1967, thanks to the energetic initiative of Florey, the Society moved to 6 Carlton House Terrace—which at least had the advantage of having the Athenaeum Club as its immediate neighbour! Robinson was very fond of the Athenaeum, but had no illusions about it. When he was a member of the Committee in 1947 he wrote an amusing letter to his mother—to whom he was always a very dutiful correspondent—saying:

> This is a dreadful hotbed of intrigue—I never enter its doors without finding some backstairs way of solving problem that ought to be tackled frontally. E.g. I saw Sir Alan Barlow [second Secretary at the Treasury] and fixed up the representation on a very important Committee and soon I ran into two other men, one of whom wants to square me and the other I want to square myself. In fact without this Club the good old B. Empire would be quite unworkable so far as I can judge.[5]

Assessing in 1979 the pros and cos of the move to Carlton House Terrace, Hodgkin thought:

> the new arrangements worked out by Lord Florey for the Royal Society at Carlton House Terrace and for the scientific societies housed in its old premises [Burlington House] represent a satisfactory compromise between

Sir Robert Robinson's grand vision and Lord Adrian's feeling for historical tradition and independence from government.[3]

This was not Robinson's view, however: for the rest of his life he regarded the new premises[6] as a costly mistake. In the second (unpublished) part of his *Memoirs* he claimed that some of the hesitant scientific institutions—notably the Medical Research Council and the Linnean Society—had been won round to his view, as well as nearly all of the leading scientists who had been consulted. The only out-and-out opponent of whom he was aware was Sir George Thomson. He also thought that Lord Salisbury, as Lord President of the Council, was hostile, but later discovered that this was incorrect. He believed that he might have realized his ambition had he formally launched the scheme a year earlier, thus giving himself longer to promote it. Of Carlton House Terrace he wrote:

> . . . here, the Society certainly has a dignified home but not one, in my opinion, at all adequate. The officers and visitors to the Society have excellent quarters but it cannot be said that the accommodation for the Society as a whole is so much increased, compared to that which was available at Burlington House, as to have made the change really worthwhile. The house is somewhat awkward with its various levels and with no really characteristic architectural plan. I do not believe that the Society will stay there very long . . . I write nearly half a century after the abandonment of the scheme [for a Science Centre]. The site is still unused and one would imagine, therefore, that with suitable legislation the opportunity still exists. Is it too much to hope that, even after so much delay, there might arise someone with sufficient imagination to grasp the advantages for the City of London and for the British Isles, that would accrue from the realization of our dreams.[7]

Bearing in mind Britain's economic problems immediately after the war, and the general atmosphere of austerity, it is very debatable whether Robinson's dream could have been realized during his Presidency. However it might well have been realized as an alternative to 6 Carlton House Terrace had his successors been minded to put their weight behind it.

Two major events during Robinson's first year of office emphasized the need for more spacious accommodation of some kind. One was the Newton Tercentenary Celebration, due to have been held in 1942, but postponed because of the war, and the other the allied British Empire Scientific Conference. Both brought large numbers of scientists from all parts of the world, the first occasion on which this had been possible since 1939. The Newton Tercentenary was significant in that it brought to London a substantial Russian delegation. This reciprocated a visit by a score of eminent British scientists—ten from the Royal Society—in the summer of 1945 as guests of the Soviet Academy of Science, on the occasion of the celebration of their 220th anniversary. Robinson had been leader of this and recalled going to a dinner at which there was 'a gentleman with white hair whom I was told was Stalin'. His doubt arose from the fact that Moscow was plastered with posters of Stalin having a shock of dark hair: Robinson remarks that 'nature will take its course' even with the great.

Another important task was the reactivation of the International Council of Scientific Unions. This had been established in 1918 to facilitate international co-operation in scientific research. All the leading countries had their own national unions, each representing one of about twenty different scientific disciplines—such as crystallography, biophysics, and pharmacology—collaborating through the International Council. As the Royal Society was the national academy of science for the United Kingdom, the British committees met under its aegis. The war had, of course, put an end to all this, and Robinson devoted much time to rebuilding the organization.

One of his last major engagements as President was to visit India in January 1950 to represent the Society at the opening of the National Chemical Laboratory in Poona by Pandit Nehru. This was timed to coincide with the opening of the Indian Science Congress. At home, just before the completion of his Presidency, he wrote a letter to *The Times* launching a public appeal for £100 000 for a Rutherford Memorial.

In the middle of his term of office, in 1947, Robinson received the ultimate accolade, for which he must long have hoped—the Nobel Prize for Chemistry. Introducing him at the royal prize-giving in Stockholm the distinguished Swedish chemist Professor Arne Fredga said that 'Among organic chemists you are today acknowledged as a leader and a teacher, second to none.' As the theme of his lecture, obligatory for all Laureates, he chose polycyclic natural products, and gave a brilliant, wide-ranging discourse—though how far his actual delivery accorded with the printed version[8] must be a matter for speculation. He concluded with a warm tribute to all his co-workers:

I conclude with an expression of deepest gratitude and appreciation to all my numerous co-workers; any success which I have had was due to their unsparing efforts. Though it might be invidious to mention individuals, yet I may be allowed to say how much I owe to the constant help of my wife, not quite my first, but much my most consistent collaborator, and over the longest period of years.

By way of light relief the Stockholm students elected him a member of the Order of the Frog.

As President, Robinson had, of course, a special responsibility, but was able to delegate much to his honorary officers and vice-presidents, and was always, as was customary, punctilious in acknowledging their support in the course of the St Andrew's Day celebrations. He had also to acknowledge the help of the permanent staff under the chief executive, at that time designated Assistant Secretary (now Executive Secretary). In the case of the latter he was particularly fortunate during the second half of his Presidency when Dr David (later Sir David) Martin was appointed to this office. At the outbreak of war he had been Assistant Secretary of the Royal Society of Arts; but then, as a qualified chemist, was seconded to the Ministry of Supply, and in this capacity attracted Robinson's particular attention. When the war ended he did not go back to the Royal Society of Arts, but became General Secretary of the Chemical Society. When J. D. G. Davies retired as Assistant Secretary of the Royal Society in 1947, Martin was an obvious choice as his successor. He was to prove a tower of strength to Robinson and successive Presidents, and to the Society generally, until his untimely death in 1976.

As President, Robinson was either host or guest at innumerable lunches and dinners; but occasions which gave him particular pleasure were the meetings of the Royal Society Club, founded in the latter part of the eighteenth century. In a sense, it was a club within a club, consisting of a select group of Fellows, who elected a few new members each year. Robinson always had some doubts about the propriety of this hyperselectivity; but none the less enjoyed the occasions enormously. The Club invited distinguished visitors to dine with them at the Athenaeum Club, and afterwards there was general discussion of topics of contemporary scientific interest. Somewhat similar, though primarily for chemists, was The Catalyst Dining Club, in the founding of which H. E. Armstrong had taken a prominent role. When Robinson first became a member it used to meet in the Ivy Restaurant, but after the war, as numbers grew, the venue would sometimes be a livery company hall. Again there were many guests, of whom the chief was asked to give an after-dinner speech with some sort of chemical flavour: this led on to general discussion.

Membership of these clubs was not, of course, *ex officio*: once elected membership was lifelong. But in his (unpublished) *Memoirs* he refers to some honours conferred on him only for the duration of his Presidency. In light vein, he mentions his election as a Knight of Mark Twain—an honour arbitrarily dispensed by Cyril Clemens, editor of *The Mark Twain Journal* and a kinsman of the famous novelist. He was also elected ex officio an honorary member of the Samuel Pepys Club, Pepys having been one of the earliest Fellows (1664). At a much higher level he was conscious that many of the honours accorded him—medals, honorary degrees, membership of foreign societies, etc.—were, at least in part, recognition of his status as President:

> On election to the Presidency there was set up, at the same time, a hat and coat rack, in order to accommodate the various honours which are bestowed on one, not so much personally, as a recognition of representation of British science. One of the more difficult tasks of the President of the Royal Society is to avoid the seductive suggestion that he, personally, deserves all this. He should know that all these things are, to a large extent, perquisites of his office.[9]

In 1950 Robinson completed his five-year term of office and, no doubt with relief, handed his badge of office to his successor, Adrian. Because of the abnormal circumstances of the post-war years it is difficult to assess his stewardship. Much time had to be devoted to rehabilitation, and in retrospect it looks extremely unlikely that the Science Centre would have materialized in this time, even without the considerable opposition to it that arose. Had it done he would, for better or for worse, have left a lasting mark on the Society's history. Probably, it would have been for better, because the Society's influence was on the wane. The government was increasingly turning for scientific advice to the very bodies from which Adrian wanted to disassociate it. On the principle 'if you can't beat them, join them', it would perhaps have been better to have rubbed shoulders on common ground.

As a most cursory inspection of the immediate post-war Fellowship and new elections clearly indicate, the Society was really representative only of academic science, whereas after the war there was rapidly growing interest in the practical application of science. The first explicitly to spell this out as President was Lord Florey (1960–65), who said:

> . . . it is the technologist who is transforming at least the outward trappings of modern civilization and no hard and fast line can or should be drawn between those who apply science, and in the process make discoveries, and those who pursue what is sometimes called basic science.

This view was accepted by Lord Blackett, Florey's successor as President, but he had had other ambitions for the Society at the end of the war. Then it was hoped not only that the Parliamentary grant-in-aid would be very substantially increased, but that the Society would determine government expenditure on the very expensive instruments, such as radio telescopes and particle accelerators, which were increasingly being sought. It is perhaps fair to say that Robinson restored the Society to its pre-war situation, but had to leave to his successors the painful task of adjusting an ivory-tower outlook to the realities of a new scientific and political world.

Although a director, Robinson took no active part in the family business. Nevertheless, through a chain of circumstances so curious as to be worth recounting, his Presidency of the Royal Society enabled them to add a profitable new product to their already considerable range. Presiding at a Royal Society dinner in 1949, Robinson had as guest of honour on his right Sir Vincent de Ferranti, President of the Institution of Electrical Engineers. In the course of casual conversation Robinson remarked quite inconsequentially that his family had made a great deal of money by making cheap disposable sanitary towels. Unexpectedly, Ferranti expressed the greatest interest in this: it transpired that his daughter and son-in-law—with a young family and no domestic help—had developed a disposable baby's nappy, but had found no manufacturer prepared to market it. Robinson promptly gave him Victor's telephone number, which Ferranti conveyed to his daughter, Mrs Valerie Hunter Gordon, in Malta that same night: she telephoned Victor the following morning. Agreement was reached very quickly, and thus was born Paddi-pads, a blessing to generations of young mothers. Recording this episode Col. P. Hunter Gordon remarked:

> It just shows how useful it is to have the Presidents of the Royal Society and of the Institution of Electrical Engineers in your families.[10]

The end of his term of office left him free to devote more time to his duties at the D.P., but before returning to the Oxford scene it is appropriate to say something of his subsequent presidencies of two other distinguished bodies.

In 1954 he accepted an invitation to serve as President of the British Association for the Advancement of Science for its annual meeting in Bristol in 1955, a few weeks after he retired from Oxford on 31 July. Considered historically, there was a certain irony in the situation, for the BA had been founded in 1831 by a group of scientists who were disenchanted with the dilettante attitude of the Royal Society in general, and, in particular, the election of the Duke of Sussex as President, rather than the astronomer J. F. W. Herschel. Frustrated in their attempts at reform—which, as we have seen, were not achieved until 1847—they established a peripatetic society which would meet annually in different cities throughout the United Kingdom, initially in York. Free from the stifling élitist atmosphere of London society, scientists would talk not only to each other but to members of the general public. Predictably, the new body was derided by those from whom it had broken away—Charles Dickens, who should have known better, satirically described it as 'The Mudfog Association for the Advancement of Everything'—but it quickly established itself as an important forum for the discussion of the dominant scientific topics of the day. The USA founded a similar body in 1848, and scientists in many other countries—India, Australia, Italy, among others—followed suit. The meetings were sectionalized according to discipline—such as chemistry, physics, geology—but there were some plenary sessions, and the Presidential Address was the highlight of the meeting, at which attendance was commonly numbered in thousands. Its proceedings were also very widely reported in the press. The Oxford meeting of 1860 attracted enormous interest though the famous clash between Bishop (Soapy Sam) Wilberforce and T. H. Huxley, the leading champion of Darwin's heretical theory of organic evolution. The 1897 meeting was famous for the warning Sir William Crookes issued that the world faced starvation if new kinds of nitrogenous fertilizers were not found. Within Robinson's own experience, the standing of the BA had been demonstrated by the influence its meeting in Australia in 1914 had had on his appointment at Liverpool.

In the immediate post-war years the BA still followed a programme appropriate to the more leisurely pace of earlier days.

Meetings lasted from mid-week to mid-week, and included visits to many local factories and the surrounding countryside, as well as a formal service in the local cathedral on Sunday. The local University provided facilities, especially lecture rooms for sectional meetings, and the city visited organized a formal civic reception.

As always, the main event in Bristol was the Presidential Address, delivered in the Victoria Rooms. The occasion was colourful to a degree: the university present in their full academic robes, and the Lord Mayor and Corporation of the ancient city in their traditional dress.

It was to this gathering that Robinson addressed himself. His title seemed non-controversial to a degree—'Science and the Scientist'—and his main theme was the desirability of freer exchange of information in the then relatively new field of nuclear energy. As so often happens, however, the main text was ignored, and a single point—little more than an aside—was seized upon by the Press. He remarked that an increasing scale of atomic explosives might eventually involve us in a 'conflagration', and wondered whether our knowledge of atomic fission and synthesis was sufficiently complete to ensure us against such a catastrophe. The fact that he was not an atomic physicist but an organic chemist, who had, as noted earlier, recoiled from the mysteries of quantum theory, was neither here nor there: the words were those of a great scientist. There was a considerable furore, and leading physicists were drawn in to express their views. They included not only local professors at Bristol—C. F. Powell and M. H. L. Pryce—but Nevill Mott, Cavendish Professor at Cambridge, holder of the chair once occupied by Rutherford, the father of atomic physics. The situation was complicated by the discovery of substantial discrepancies between the pre-published speech and what he actually said. This was thought by some to have sinister connotations, but was in truth no more than a consequence of the fact that he was over-running his allotted time—which on a less ceremonial occasion would not have worried him at all—and he had to cut his text as he went along. The physicists were naturally loath to discomfit him by a flat denial, and took refuge in bland phrases such as 'never being 100 per cent sure' (Pryce); 'unlikely' (Powell); or 'so unlikely that I should say they are zero' (Mott). However, this slight prevarication enabled Robinson to claim that he had done no more than draw attention to a possible 'loophole'. Unperturbed, he developed the point further in an address to the local Rotary Club, warning that nuclear power on a large scale involved hazards then still not properly understood, and maintaining that reserves of fossil fuels were still greater than many supposed. As a point of a local interest, he argued that one-seventh of the country's electricity needs could be generated by a Severn Barrage—still, more than thirty years later, only at the stage of very detailed evaluation. In the post-OPEC era, and with the strong environmentalist lobby against nuclear power, the reports of this address today read rather well.

Only in one other respect did Robinson cause any anxiety to the officers of the BA, and that was over the catheral service. This was an integral part of the programme, and as such it was mandatory that the President should attend. Robinson, however, had no time for organized religion of any kind, and flatly refused to go. It took much coaxing to persuade him that no sacrifice of principle was involved: he would be present as President of the BA, and not in any personal capacity.

During his visit to Bristol, Robinson stayed at the Mansion House as guest of the Lord Mayor. At that time the latter was chairman of the Kleen-e-zee Company, a thriving local firm which had built up a considerable business by the door-to-door sale of brushes, dusters, polishes, and other household cleaning materials. They apparently got on very well together, and Robinson repaid his hospitality by enlightening his host on the advantages of incorporating silicones in polishes as a waterproofing agent.

Among the incidental duties he had to perform was to present prizes to the winners of an annual scientific essay competition sponsored by *Endeavour*, of which I was then editor. There was an age-limit of twenty-five, and I remember very well how easily he put these young scientists at their ease, taking a real interest in the work they were doing and their ambitions for the future.

The other society of which Robinson was President was the Society of Chemical Industry. Relatively speaking, this was a newcomer, having been founded in 1881 to serve the interest of the growing number of chemists who were employed in industry. Not surprisingly, therefore, it had strong affiliations with Manchester and the other northern centres of the chemical industry. The first President was H. E. Roscoe, and two of his Vice-Presidents were W. H. Perkin and E. K. Muspratt. As the Royal Society alternated the Presidency between the A and B sciences, so the SCI alternated between academic and industrial chemistry: although Robinson was elected on the former ticket, he had, of course, far more industrial knowledge and experience than most university chemists.

Like the BA, the SCI held its annual meeting at different centres each year, very occasionally venturing abroad. In his term of office, 1958/9, he devoted his Presidential Address to 'Organic Chemical Research in Industry'. It was clearly an uneventful, though obviously not unimportant, year, for he records in his unpublished *Memoirs* simply that:

> In 1958–1959 I was President of the Society of Chemical Industry. I have nothing of general interest to report though considerable progress was made in detail.[11]

According to the terms of his appointment at Oxford, Robinson should have retired in 1951, when he reached the age of 65. In the event, he continued for another four years; and how this came about is of some interest. The award of the Nobel Prize in 1947, when he was just turned sixty, was a considerable stimulus, and he was inclined to take on more work rather than less, and the thought of retirement was irksome. He once discussed this with Haworth, who was in a similar situation at Birmingham. They each decided to make formal application for an extension, on the rather specious grounds that two world wars had effectively robbed them of ten years of academic life—specious in the sense that the same could be said of countless academics. Haworth failed at Birmingham, but Oxford unexpectedly acceded to Robinson's request, and he continued until just before his 69th birthday, thus occupying the Waynflete Chair for twenty-five years. Interestingly, his colleague at the D.P., D. Ll. Hammick, also applied in respect of his praelectorship, but was turned down. He was not surprised, because he had already confided his doubts to a colleague, saying 'He'll get it and I won't—sheer eminence will get it for him'.[12] However, Hammick was awarded a consolation prize, in the form of an appointment as Assistant Professor.

This extended tenure was productive in research papers—some sixty in all—but, sadly, it ended in grief. In 1953 the University of Oxford recognized Gertrude's unique qualities by conferring on her an honorary MA degree. In presenting her for the degree the Public Orator—speaking, of course, in Latin—succinctly summarized the way in which she had complemented Robert:

Nowadays, when parents themselves do so much of the work once given to domestic servants, I often reflect on the great debt we owe to the wives of men distinguished in their profession. Preyed upon by numerous household anxieties they still have a smiling face and pleasant words of welcome for earnest savants coming from any country; and after hospitable entertainment send them on their way with every show of extreme reluctance. Here is one such wife, married to 'A mighty man of mind' and deserving even more praise than most; for she shares his professional interests and has not merely relieved him of many domestic cares but helped him in his laboratory work, both by co-operation and by her own discoveries—adding, incidentally, a subtle flavour of informality and charm to that high seriousness which chemical study demands.

She enjoyed this honour only briefly, however, for she died in the following year after a heart attack.

It was a grief to Robert that he was not with her at the time. She had been unwell, but not so much so as to make him feel obliged to cancel a visit to Holland. He was greeted with the news when he reached Harwich on his way home from a visit to Karl Ziegler at Mulheim-in-Rhur. His grief was sharerd by a vast circle of friends and colleagues: in all he received almost a thousand letters of condolence.[13] To these were added more formal tributes in the shape of published obituaries, all emphasizing how effectively she had complemented Robert as his career unfolded. Thus John Simonsen, their old friend from student days at Manchester, wrote:

She shared to the full her husband's scientific and other interests. She was a skilled climber and travelled widely with him. Any account of her life would be incomplete without a reference to her social gifts. She was a born hostess. Those who had the privilege of being her fellow students in Manchester will still recall with pleasure her invitations to dances at Ashbourne Hall, the Women's Hall of Residence. Later, in the succession of universities in which her husband was professor, she took the deepest interest in the welfare of the wives and children of his staff. She seemed never to forget a name or a face.

The full opportunity for her to exercise her great gifts came, however, in Oxford. There, in her beautiful home or at Magdalen College, she showed gracious hospitality both to Oxford friends and to the numerous visitors from home or overseas. No scientific conference in Oxford was complete without one of her delightful parties.[14]

Again, Wilson Baker, who had also known them at Manchester, but more particularly at the D.P., wrote:

Her contributions to chemical knowledge have been substantial in spite of many pressing calls on her time and energy . . . it is in the dual role of an indefatigable research worker and a kindly hostess that she will long be remembered. In particular, the younger members of the teaching staffs and their wives have reason to be profoundly grateful to her; she found time to take a personal interest in them and their families, which lasted long after they moved elsewhere.[15]

The funeral was held at the local cemetery in Wolvercote. So many mourners attended that the accommodation in the chapel was quite inadequate, and many had to stand in the aisles and the entrance.

Although the Robinsons had continued to work in the same laboratory, and constantly discussed each other's research problems, their professional paths had in fact been divergent for a number of years. Their last joint publication, on flower pigments, was in 1939; and during the war they did some joint work on penicillin. Thereafter she gave particular attention to the isolation of a growth-inhibiting substance, with T. B. Heaton. Little of this was published, however, save for a couple of brief contributions to *Nature* in 1943 and 1948.

Thus 1954 was a black year. With Gertrude gone and the end of his professorship imminent, the future must have seemed bleak and unpromising. Nevertheless, it was in fact the prelude to a new era, both personally and professionally.

11

An active retirement

After Gertrude's death circumstances were such that Robinson could have no direct support from his two children. Michael—mentally handicapped from birth—was a permanent and worrying liability and Marion's career had followed a course very different from anything her parents had imagined, and one to which he was not, indeed, particularly sympathetic. The general ethos of Headington School helped to develop strong religions convictions, and as a girl she decided to dedicate her life to furthering Christian ideals. After graduating at Oxford from Lady Margaret Hall in 1943, she qualified in medicine at University College Hospital. She then told her parents that she wished to take a diploma in child health—which she did at the Royal Victoria Hospital, Newcastle—and followed this by a Diploma in Tropical Medicine and Hygiene at Liverpool. Then, in 1949, she dropped what Robinson[1] described as a 'bombshell' by announcing that she intended to work in Africa as a medical missionary. In January 1950 she took up an appointment as Medical Officer at the UMCA Hospital, Lulindi, Masasi, in the Southern Province of Tanganyika (now Tanzania). With the limited resources available, versatility was essential, and she had to apply herself to several fields, including surgery, X-ray diagnosis, and the training of nurses. Though she was anxious to return home, she had to honour existing commitments, and she did not get back until the autumn of 1957, initially to spend a short time in private practice. But by then, as we shall see, Robert's personal life had taken a dramatic new course.

But he had, of course, a wide circle of Chesterfield relations. His mother, to whom he was closely attached, had died after a very long illness in 1950 at the age of eighty-six, but all his younger siblings survived. He was particularly close to Victor, who had been appointed Chairman of Robinson and Son Limited in 1945, retiring in 1961. His eldest sister Florence had been elected a director in 1939, and was very busy in local affairs. She served as Mayor in

1946, and during her term of office Cecily, who had come home after internment in Hong Kong, acted as Mayoress. Robert, of course, played no part in such civic activities, but he always took particular pride in the fact that Chesterfield elected him a Freeman in 1947. Only his youngest sister, Dorothy (Dolly), was close at hand. She, it will be recalled, had married Alex Bell, who was in charge of the London office of Robinsons. This was severely damaged by a direct hit during the blitz, as was their home nearby. After this disaster she returned to Chesterfield, and lived there for a time. Then they bought an old manor house at Great Milton, a dozen miles east of Oxford. The intention was partly to start a private hotel, and partly to set up their son David as farm manager. Unfortunately, Alex's health failed, and he had to retire in 1953.

After Gertrude's death Robert soldiered on as best he could for the remainder of his term at the D.P., but, scarcely surprisingly, he became increasingly unpredictable. J. C. Smith has recalled the awkwardness of the situation:

> I, for one, avoided his office unless urgent departmental business forced the visit. Of course he could be charming and helpful, but more likely one would find oneself involved in a bitter argument about such a matter as the high cost of analyses, or as to why someone had refused a job which RR thought suitable. The real business had to be postponed or dropped.[2]

Nevertheless, despite the inevitable disruption he remained as busy as ever, as is indicated in a letter in 1955 to David Ginsburg, who had by then moved on from the Weizmann Institute to the Technion in Haifa, and invited him to pay a visit there:

> Many thanks for your kind invitation, but this time I cannot possibly manage to attend. I have to be in London on June 23, 24, 31 and July 1st and in Manchester on July 8 and 9. Immediately I go to Lindau then to Caen (Association Française) etc, Switzerland and back to vacate my house in Oxford, which I am selling. Then in September the British Assoc. etc. Added to all this July 31st is my last day at the University and I have deep roots to pull up.

The move from Banbury Road—with all its associations and the lovingly tended garden—must have been a great wrench. In its place he bought an attractive house, Grimms Hill Lodge, on the hillside above Great Missenden, and found a good housekeeper to look after him there. In those days it was an easy drive to Egham, some twenty-five miles away, and Heathrow. The commuter service to

London was good, but apart from that he had also a flat in Mount Street, in the West End.

As his retirement became imminent, he was asked if he would sit for a portrait to be hung in the D.P., but he refused. Instead, he asked for an album of signed photographs of all his colleagues and research students. Inquiry showed that this was more easily said than done, for more than 240 people were involved, they were scattered all over the world, and no records of their whereabouts had been kept. Undaunted, Muriel Tomlinson set about the task systematically, and after writing some 600 letters gathered together 214 photographs. The result clearly pleased Robinson greatly. The presentation was made at a farewell dinner in Lincoln College on 4 October 1955, and J. C. Smith—despite all their differences—gracefully made him a personal presentation. This was a dinner plate boldly inscribed with the initials RR: this, it turned out, he had acquired from the Rhodesian Railway during a visit to Victoria Falls a few months earlier! Although the D.P. was deprived of an oil painting, they did acquire a good portrait photograph. Despite the special dispensation to extend his professorship, the only such extension ever authorized by the University, Robinson still hoped that some means might be found to enable him to continue his research in the D.P. Early in 1954 he had written to Sir Douglas Veale, the University Registrar, urging him to organize the appointment of a successor as soon as possible 'to facilitate his own arrangements'. The post was duly advertised, at a salary of £1900 plus £400 as head of department.

In the event, the electoral board, which under the normal procedures of the University was not required to limit its consideration to people who had applied for the post, decided to invite E. R. H. Jones (now Sir Ewart Jones), who had for the previous eight years occupied the prestigious chair of organic chemistry at Manchester, to accept the appointment. He had been elected FRS in 1950 and, like Robinson, had a particular interest in the structure, synthesis, and biogenesis of natural products. After careful consideration Jones agreed in principle. Robinson was well pleased, telling Veale that he thought this much the best choice that could be made. Nevertheless the circumstances of the appointment were such that Jones had to make certain important stipulations before committing himself formally. On the one hand, there were practical considerations arising from uprooting himself and his group in Manchester and transplanting it to Oxford. Much more serious, however, were problems arising from the state of the laboratory which he would inherit. Partly because of the after-effects of the war, partly because the University then had no proper organization for survey and maintenance, but largely because Robinson had for years been so much occupied with external affairs, the whole building was very dilapidated. It was over-crowded, and the standard of cleanliness was low. It was then some thirty-five years old, and the wiring was all rubber-insulated, making it much less durable than modern pvc insulation: at some points bare wires were running parallel to one another. There was a complex network of gas pipes, mainly to serve the bunsen burners of the individual students and research workers. Much of this was rusted, and at some points within walls proved to be leaking dangerously. The facilities for storing inflammable solvents were quite inadequate. All this would have horrified modern health and safety inspectors, though it must be recognized that thirty years ago attitudes throughout the country were generally pretty casual. Laboratory facilities generally were outdated: the 1940 extension, for example lacked a cold room, a fire-proof room, and glass-blowing facilities. Another serious deficiency which came to light only later was the accumulated collection of organic chemical samples—some dating back to Perkin's day—which Robinson carried off with him. Some were eventually recovered, but in too confused a state to be useful. Robinson was among the last of the great chemists in the classical tradition, relying on very simple apparatus—test-tubes, flasks, condensers, waterbaths, balances, and so on. With this he and his colleagues obtained remarkable results, but the chemical world generally was moving on towards much more sophisticated and expensive equipment, such as infrared and ultraviolent spectrometers, mass spectrometers, and gas chromatography. These deficiencies at Oxford were well known among organic chemists, which may explain why there were so few applicants for the chair when it was advertised. Robinson seems to have been unaware of this, as he told Veale that this was 'the foremost chair in Organic Chemistry in the country'.

Jones was confident that with proper support the D.P. could be put on its feet again within three or four years, but was not prepared to commit himself unless this support was assured. After due deliberation, the University found it possible to promise to meet his main requests without undue delay. A third-floor extension, accommodating ten research workers, could be started in 1957, following an already agreed extension to Human Anatomy. Additionally, an adjacent site—the only one still unallocated in the science area, and originally intended for the Mathematical Institute—would be earmarked for a new building altogether. Some funds, and the prospect of more, could be made available for the purchase of equipment and apparatus. On this basis the business was concluded to the satisfaction of all concerned, and Jones formally accepted the appointment.[3] Despite the University's special dispensation to extend the term of Robinson's professorship, he cherished hopes that even after his retirement he might be able to continue his own research with a team of eight in the D.P. With this in mind he asked Professor Jones if he could find some space for him. Jones had to decline, giving the strictly practical reason that the laboratory was already overcrowded, and that he would be bringing with him sixteen of his own research workers from Manchester, as well as his laboratory superintendent. Apart from the impracticality, he was surely right. It is never easy for a new incumbent in any position to have his immediate predecessor adjacent under the same roof—and the more so when that predecessor was as unpredictable as Robinson. It was, in any case, the Oxford tradition that a retired professor did not linger in his own laboratory, though some well-disposed colleague might offer him facilities elsewhere. Thus the link with the D.P. was finally severed, though he continued as an Honorary Fellow of Magdalen, and the University awarded him an honorary degree at Encaenia in 1956. There we, too, must leave the D.P., but not without noting that Sir Ewart Jones did indeed, fulfil his promise to restore its faded fabric.

Although he failed in his bid to continue his research at the D.P., retirement in no way left time hanging on Robinson's hands. Apart from the immediacy of the Presidency of the British Association, there clearly stretched ahead an almost indefinite series of activities in the academic sphere—attendance in one capacity or another at conferences at home or abroad, awards of honorary degrees, and so on. But, additionally, a combination of circumstances opened up quite a new sphere of activity as the doors of the D.P. finally closed behind him. As has been noted earlier, he had been associated with ICI as a senior consultant since its formation in 1926: indeed his connection was even longer, as he had for a short time been Director of Research in the British Dyestuffs Corporation, one of the initial constituents of ICI. From 1939 his connection was primarily through the Dyestuffs Division Research Panel: the two other external members were Alexander Todd (now Lord Todd) and Heilbron, who resigned in 1949 when appointed the first Director of the Brewing Industry Research Foundation. As all three were well-known to each other, both personally and professionally, it was an agreeable as well as an effective arrangement. During these years ICI had had an interest in making oil by the hydrogenation of coal by the Bergius process.[4] In the 1930s this was not viable commercially without a measure of tariff protection, given by the government in 1933: during the Second World War it produced large quantities of urgently needed aviation spirit. After the war economic factors were again unfavourable, and ICI finally closed its Billingham plant in 1958.

Although this development had strong economic and political implications, it was of interest to Robinson in enlarging his already enormous scope of chemical knowledge. At that time the British chemical industry was almost entirely based on coal-tar as a raw material. This was in marked contrast to practice in America, where petroleum—abundantly available from indigenous sources—had been used as a raw material on an increasing scale since the 1920s. This led to the development of various 'cracking' processes, by which the long-chain molecules in petroleum could be broken down into shorter ones, useful both as motor fuel and for chemical synthesis. In this an active interest had been taken by Robinson's old friend Chaim Weizmann, with whom he had kept in regular touch since their Manchester days. At the Daniel Sieff Laboratory (later incorporated in the Weizmann Institute at Rehovoth) Weizmann and E. D. Bergmann developed what was called the Catarole cracking process. This attracted a good deal of interest at the time, and in Britain a company known as Petrochemicals was set up to exploit it, largely at the instance of Georg Tugendhat, a Viennese industrial chemical consultant, who also founded the Manchester Oil Refinery Group. Much of the capital came through a government agency, the Finance Corporation for Industry (FCI). To watch its interests, the government reserved the right to nominate two directors on the board, and of these Robinson was one. This was a convenient arrangement for him, as the manufacturing plant was at Partington, near Manchester: he could thus easily combine visits there with those to ICI's Dyestuffs Division Research Panel at

Blackley. He also became a director of Petrocarbon, an associated company concerned with research and development. Meanwhile, ICI had been developing their own interests. In 1945 they recognized that in their post-war development the best source of raw materials would be bound to be petroleum, and that they would have to build appropriate cracking plants, particularly after the government withdrew the import duty on hydrocarbon oils in 1946. This led ultimately to the development of a new ICI site at Wilton, where the first cracking plant came into operation in July 1951.[5]

Later, Petrochemicals found itself in difficulties because the Catarole process, its mainstay, had been rendered obsolete. With his foot in both camps, Robinson hoped that it might have been acquired by ICI, especially as the asking price was not high, but to his disappointment they declined.[6] No doubt, however, there were good reasons. On the road to Wilton, ICI had had complex discussions with Anglo-Iranian and the Distillers Company about joint enterprises, but nothing eventually came of them, and they had clearly decided to pursue their own course. Indeed, with the Catarole process obsolete, Petrochemicals had seemingly little to offer except a developed site and some experienced industrial chemical staff, for neither of which ICI had any great need at that time. In the event Petrochemicals was acquired by Royal Dutch Shell, whose needs were rather different. Although they had from the 1920s developed large chemical enterprises in the USA, mainly to utilize surplus gas from their vast oilfields, their chemical operations in Europe dated effectively only from 1943, when a large plant—commenced in 1939—for the manufacture of detergents and wetting agents was opened at Stanlow, near Liverpool. In 1949 this was extended to the manufacture of alcohols and solvents. In the same year they opened other works in Rotterdam, to manufacture pvc plastic, and in Rouen. This was the start of a great expansion programme, and they were naturally in a much more acquisitive mood than a long-established and rather staid chemical company such as ICI. With Petrochemicals, Shell also acquired Robinson, who severed his 35-year-old consultancy arrangements with ICI and its predecessors in order to join them. He became a director of Shell Chemicals UK Ltd, and, from 1967, of Shell Research Ltd, holding both these offices until the time of his death. By a special dispensation, Shell in 1960 allowed him also to act as a consultant to Beecham, to advise them on penicillin and its semi-synthetic analogues. Thus once again the wheel had turned, and he had gone back to play an active role in the chemical industrial world. At that juncture in his life it was a most timely development. Perhaps most important, Shell provided him with research facilities on a whole floor of the Shell Technical Service Laboratory at Egham. It had to be completely re-equipped to meet his requirements, and he spent on average a day a week with his collaborators there. He was fortunate in persuading Dr Renée Jaeger, who had been with him at the D.P. for many years, to go with him as deputy manager of the Egham Laboratory. He also took with him G. I. Fray, who had been doing a D.Phil. at Oxford under N. Polgar. (He stayed there until 1962, when he moved to the University of Bristol.) Fray recalls that unless he was visiting elsewhere Robinson visited Egham weekly, and spent an hour or two with his research team, holding a colloquium at the blackboard. Although he would sometimes sit with closed eyes, he was far from somnolent:

> We sometimes thought that he had actually fallen asleep, but woe to the speaker who strayed from the paths of sound chemistry! His eyes would open at the first incautious remark and his retort could be devastating.[7]

When he switched his mental concentration to full power, 'his eyes appeared to darken as the pupils dilated'. Fray, like others, remembers the two sides to Robinson' nature, how at times he could be generous with praise:

> I once had a rare flash of inspiration concerning a very minor structural problem and started to explain. Before I had finished RR said 'That's right!', turned away with his characteristic fierce expression, then suddenly swung round with his equally characteristic beaming smile and pronounced the idea brilliant. His kindness to young members of his research team included occasional invitations to dinner: I remember accompanying him to a Royal Society dinner at the Athenaeum and to a Catalyst Club dinner.

This arrangement continued until 1973, when the whole site was closed as a result of a policy decision by Shell to concentrate its research activities at Sittingbourne. Apart from providing the research facilities he wanted, the job provided ample opportunities for the extensive travels he so much enjoyed. His work took him to Shell's main centres in the UK, notably the research laboratories at Thornton and Sittingbourne, as well as others in Amsterdam and at Emeryville and Modesto in California. Moreover, as an important emissary of a rich company he could travel in the style he enjoyed so much. Maurice Stacey recalls an occasion in 1951 when he travelled with him and several other distinguished chemists—including Sir Harry Melville and Sir Harold Thompson—on the *Queen Mary* to attend the 75th Anniversary Meeting of the American Chemical Society. They were all travelling tourist class, and Robinson was infuriated to encounter some very junior ICI chemists who were travelling first class. At a later meeting of the American Chemical Society in 1955 he was involved in a curious episode, when a radio report erroneously announced his death: the shock was considerable, and the members present stood briefly in silent tribute. On the following day he was glad to assure an audience in Pittsburgh that, as with Mark Twain, the account of his death had been greatly exaggerated: he had been confused with his friend Robert Robison.

But this luxury travel was to pay a much higher dividend than mere comfort. In the early 1950s the preferred way of trans-Atlantic travel was still in one of the great Cunard liners: not until 1957 did more travellers make the crossing by air than by sea. On such voyages, travelling first class, much was made of Sir Robert Robinson FRS, OM, and he was regularly accorded a seat at the captain's table. It was on one such occasion that he met the lady who was, in 1957, to become the second Lady Robinson. It was a marriage that began happily, soon became fraught almost to the point of disaster, but ultimately settled to a mutual understanding and tolerance. Robert was clearly very much attracted on first acquaintance, but not until January 1957 did he say anything, even to his brother Victor. Then, after a brief reference to Marion and his new interests at Shell, he wrote:

I was sorry not to have been able to visit you all at Xmas . . . but to be quite honest (as I desire to be) there was another factor that supervened. At this stage I think you should take a reliable shock preventive! Over about 3 months I cultivated a friendship with a charming American Lady who is Vice President of a large advertising & agency firm in U.S.A. We went to quite a number of functions dinners, theatres etc together and I visited twice her hostess at Windsor Manor, Sunningdale. Her friend was formerly Binnie Barnes a musical comedy actress and now a county lady on a large scale. All these people are very well off. Eg B.B. keeps two flats in town and her house is furnished regardless with 17 bedrooms and 7 bathrooms! Sylvia (my friend) clubbed with 3 other Americans to charter a plane, necessary as she had 8 crates of purchases and 15 pieces of luggage as well. Nothing can come of all this except that when next in N.Y. she has promised to take me to 'My Fair Lady' which is usually booked up 6 months ahead. As an example of the kind of person she is I might mention that in spare time she has devised a new character of children's Television and both N.Y. and London are taking it. Also I took her to a play and a few days later was informed that she was negotiating the American rights![8]

Reading between the lines, and knowing the outcome, one might infer from his comment that 'Nothing can come of all this' meant that he was not altogether without hope that something would. As it did, it is easier to understand the later course of events if something is said at this point about the personality and background of his new friend. This is not easy, in that there is no single source from which a comprehensive picture can be drawn, and it is therefore necessary to try to piece together various small—and sometimes conflicting—items of information gleaned from sources listed in the Preface.

Stearn Sylvia Hershey was born in Akron, Ohio, on 14 February 1911. She was of European descent; her father is variously described as an Austrian lawyer and a Hungarian real-estate manager. Her mother was of Hungarian origin. She had an older and a younger brother, to whom she felt close. Her first marriage to Nils Hillstrom of New York brought her into the world of entertainment: he was an executive of the Muzak Recording Company. They had a daughter Stephanie, born on 3 October 1936, but the marriage ended in divorce, and Stearn went for a time to live with her mother in Detroit. Then she worked as a journalist, and later as a radio scriptwriter responsible for a daily programme in Los Angeles: she spent eight years there in the days when Hollywood was at its peak. She then returned to New York as Vice-President of a large advertising agency, which also had interests in plastics, but continued her freelance activities. Her experience in Hollywood—and contact with leading figures there such as the actress Binnie Barnes (whose house at Sunningdale had so much impressed Robinson) and the film director Mike Mankowicz—led her to take an interest in motion pictures as a script writer.

Such, in brief, was her career up to the time of that fateful meeting on the *Queen Mary*: she was successful, and correspondingly rich, in a business where big money, and talk of big money, was the order of the day. She had a personality to match: outspoken (with one of the less attractive American accents), self-confident, self-assertive, extrovert. Seemingly these were qualities which Robert found attractive, for their acquaintance on the *Queen Mary* began with a heated argument about Red China, in the course of which Stearn told him that 'he was shouting out of his mouth from both sides'. That he even tolerated this is remarkable, considering how ill he took the slightest contradiction from chemical colleagues. She was also a gambler, 'ready to bet on two flies climbing up a window pane', as one close friend put it.

It is not surprising that such a person was received with mixed feelings by Robert's existing circle of friends and colleagues, whose social patterns tended to be shaped by the standards of the Royal Society, the Athenaeum, and the Common Rooms of the Universities. It was as though some sort of Zuleika Dobson had come among them. This unorthodoxy tended to colour their reaction, which was not always favourable. Yet there is every reason to suppose that Robinson himself thoroughly enjoyed this new world. Among friends with her own background she was clearly well like. Thus Helmut Meyer of New York, her literary agent for many years, wrote:

> I can state, without reservation, that Stearn was one of the most remarkable and many-sided talented women: not only spirited and beautiful but endowed with a terrific sense of humour . . .[9]

Roald Dahl, the writer, who came to know her well later as a neighbour at Great Missenden, remembers her as 'nice, kind, sweet, and generous'.[10]

However one interprets the characters of both, the disparity in interests and background was clearly enormous. One of their mutual interests was a passion for the word-game Scrabble. There was also the difference of age, though this was much less than that between Robert's father and his mother. When they first met Robert had just celebrated his seventieth birthday, and Stearn was a mere forty-five. It is idle to speculate why both were eager to embark on marriage, for logic does not enter into such matters; but some factors seem clear enough. Robert was clearly flattered to have attracted the attention of a talented, much younger woman who introduced him to a new social world of whose very existence he can scarcely have known. Whatever a new marriage might hold, it would clearly be very different from that with Gertrude. At a lower level, he was lonely—for Marion was still in Africa winding up her affairs before returning home—and the thought of a settled home, and home comforts, must have weighed with him. Indeed, he said as much in a letter to his nephew George—then a District Office in Northern Rhodesia—telling him of his forthcoming marriage: 'I am tired of being alone'. While there is no reason to doubt that Stearn had a genuine affection for him, she too must have been attracted by the thought of a new kind of life with a titled and supposedly wealthy husband who travelled the world and was treated with great respect wherever he went. According to her daughter there is also reason to suppose that psychologically she found an older man reassuring.

Be this as it may, the courtship ripened swiftly: by February 1957 they had an understanding, if not a formal engagement. Robert had met Stearn's daughter Stephanie—then twenty years old and engaged to be married when she graduated in the summer of 1958—and the two had taken an immediate liking to each other, so all was well on that front. Marion, however, was less happy, as another letter from Robert to Victor makes clear:

> She is highly suspicious, I think, about my American friends and may feel it her duty to come to the rescue. My own obsession is to see her financially independent, but she says money does not interest her at all. A noble but impractical attitude, I never having experienced any such difficulty in that direction.[11]

Victor wrote by return of post in response to Robert's startling news, and although his letter does not survive it would seem from the wording of Roberts' own swift reply that he was concerned about the disparity of age:

> Thank you for your observations. What you say *re* Sylvia is of course perfectly true—we know it and both wish it could be otherwise.[11]

His next extant letter to Victor is a mixture referring only indirectly to Stearn, as usual in the same breath as inconsequential trivialities. His main concern is that the *Queen Mary* may not run to schedule:

> The QM trouble makes me feel doubtful about my joining in mid-April [but] this morning I got a request for table instructions so it looks as if they are expecting to recover their schedule. This is of *vital* importance to me as I am promised a seat at 'My Fair Lady' the fabulous NY show. Nothing further from Marion . . . Interesting about your cat. Pippa I died at 17.[12]

Whatever misgivings might have been expressed by others, and in spite of any private reservations they might have had themselves, the arrangements went on apace, and they were married at Caxton Hall on 15 July 1957. Stearn gave Binnie Barnes' home, Windsor Manor, as her English address, and there was a big reception afterwards at the Trocadero. The occasion attracted a good deal of press publicity, to which Stearn, understandably with her background, was not averse. She announced that she was forty-one (although in fact she was then forty-six) and that Robert had given her not only two engagement rings—one diamond and one ruby—but two wedding rings—'a diamond one for best and a gold and platinum one for weekdays'. He had clearly had to put his hand deep in his pocket. It was, as I recollect, a somewhat polarized event: the *News Chronicle* described it as a 'fascinating gathering'! Apart from family, Robert's guests were mostly pillars of the scientific establishment. Stearn's were, of course, very different. They included Thomas Reed, a Texas oil magnate; Mike Frankovitch, of Columbia Pictures; and Edward Gerson, who with his wife controlled the plastics cum public relations agency of which Stearn was a Vice President. Afterwards, they left on a five-week honeymoon, which was, typically, to include a week in Paris attending a chemical conference.

Somewhat to Robert's consternation Stearn had ordered a complete and expensive redecoration of the house at Great Missenden, and on their return they had to live in the London flat. Comparerd with Banbury Road the new house was relatively small, and Robert was concerned that he could not give Marion the space she wanted on her visits home. Writing to his nephew George Welsh in Northern Rhodesia in October 1957 he said:

> Talking of the house reminds me of Marion's difficulties. She wants to treat it as a *pied à terre* & a place where she can spread out her belongings and books etc & return to at any time. This is naturally not possible as it was at 117 Banbury Rd where she grew up under the indulgent eyes of fond parents . . . I am sure Marion and Sylvia will get on very well together but we have only 1½ guest rooms & clearly this has to be kept open for the occasion of guests. Actually Marion always said that she would *not* live with me. That was always an argument for not coming home. But my desire was for her to become economically independent and this still holds . . .

For the next eighteen months things seem to have gone well enough, judging from Robert's letters to Victor. He is doing much globe-trotting, on behalf of Shell and in attending major chemical conferences. A new contact is Giulio Natta, Director of the Milan Institute of Industrial Chemistry and a consultant to Montecatini, then Italy's biggest chemical company. The reason is Natta's development of a new plastic, polypropylene, akin to ICI's polyethylene. Very conveniently, Natta has a mountain house in one of the valleys south of the Matterhorn and north of Val d'Aoste. Stearn spends a good deal of time looking after her business interests in the USA and planning the forthcoming marriage of her daughter Stephanie to John Elliott (Eleuteris), a medical student who is to spend a month at a British hospital as part of his degree course. Robert likes them both: 'an attractive if somewhat uninhibited couple'. He takes them up to meet Victor and his wife Betty in Chesterfield, and the visit is a success. He and Stearn also invite them to spend a week together in the Swiss alps. John is interested in climbing, but Stephanie 'is her mother's child and has a distinct preference for funiculars. She is a surprisingly capable young woman'. Marion comes home from Africa at the end of 1957: they visit London theatres together, and Marion goes with him to Israel in the spring of 1958 to attend the 25th anniversary celebrations of the Daniel Sieff laboratory. Physically energetic as ever, he is busy in the garden at Great Missenden:

> On Sunday last I started a garden pool—it will take 7 full days to excavate—with the double object of draining the lawn and growing water plants. This is a quite unnecessary exercise but I've convinced myself that it is a good idea.[13]

So it goes on up to early April 1959, when he writes telling Victor that he and Marion had gone to see Michael in the home in which he lived at Brighton. He meets Mark Way, Bishop of Masisi, whom Marion is to marry in 1960, and who is torn between continuing to do what he conceives to be his Christian duty in Africa or returning to England. With his anti-clerical bent, Robert mischievously writes of him to Victor as 'Mark†'.

Sadly, a totally different situation is revealed in May 1959, when Stearn has just returned from a long stay in America. A violent dispute broke out: predictably, it was about money. Desperate, Robert turns to Victor for comfort and advice. Much of what passed need not be ventilated, and in any case some was done by telephone: on 11 May he wrote: 'Many thanks for your phoning; it did me good to hear your voice'. He reveals that he has been living well beyond his means, and is already dipping into capital. He has told Stearn, who, despite the fact that she could use the Mount Street flat, is living at high cost at the Westbury Hotel in London—much favoured by rich Americans—that she can return home provided she is prepared to live within his income. Nevertheless, their respective lawyers are actively exploring the possibility of divorce or a legal separation: a divorce in Nevada on the grounds of incompatibility, or, after three years, in Britain on the grounds of desertion. Meanwhile he is seeking some agreed scale of maintenance. He becomes convinced that she is mentally ill, a schizophrenic, and draws a distinction between a Sylvia personality and a Stearn personality:

If Sylvia (as distinct from Stearn) (I have indicated that I consider her a case of advanced schizophrenia) decides to return home I would be more anxious about her future. She has no conception of economy at all—I can't help fearing a wreck on the rocks of 'luxury girlmanship'. I believe my experiences would make a best-seller & I'm sure Stearn (not Sylvia) would be interested in that.[14]

Yet he is surprisingly understanding, as a letter a few days later shows:

I think Lady R is a schizophrenic. Sylvia is a sweet girl and I love her. Stearn, unfortunately is inextricably bound up in a hard business type, not much prevalent over here and brought up in the Hollywood atmosphere etc. I won't pursue it.[15]

While all this legal wrangle is going on, there are some hopes of an amicable accommodation:

Counsel will advise us (Solicitor and myself) re a possible settlement, this afternoon. Sylvia is just *too* sweet at the moment, but Stearn lies in wait behind the curtain. The S.S. complex has no sense of economy . . . I think she would like to stay at Grimms Hill until *October* but I have made it clear that I am not going back on my present happy relations with Marion.[16]

In June, he learnt that Stearn had gone so far as to file a petition for divorce, but a month later the situation had changed again, and for the better.

The situation has taken a new turn in that Stearn has withdrawn her suit and I am giving her what I promised when she first came over here in May (59). My lawyers are sure I would have won but all the same there would have been great publicity & damage all round . . . Many thanks for your letter re [Robinson] shares, I never made the promises Stearn says I did and she has *not* even now shown me the letter which refers. The question of transfer only arose in connexion with events after my decease but she evidently had a genuine misunderstanding. That I must admit and also must say she never tried to clear it all up. Our relations now are very friendly but she *is* going to

USA. Marion and I go to Switzerland end of next week. Stearn to Sweden on Th. next.[17]

Victor wrote back confirming the misunderstanding about the transfer of shares to Stearn: he had always understood that Robert's intention had been that they should be used to help support Michael and Marion after his death.

Correspondence with his nephew George Walsh nevertheless shows a miserable state of affairs continuing into the autumn of 1959:

> . . . I am terribly worried, just now and must apologize for nerves and irritability. Marion is a great comfort . . . I want to live on income and I could never do that . . . The situation is horribly complicated and self disastrous from the point of view of a quiet retirement with time to write. I have about 45 papers to write & a book to revise—two others to write *ab initio*.

The book, we must suppose, was *100 years of Synthetic Dyestuffs*, published under the Pergamon imprint.

Eventually, an uneasy truce was declared and the storm effectively blew itself out. Stearn pursued her own business interests, but spent a good deal of time with Robert either at Grimms Hill Lodge or a new, and cheaper, flat at 1 Lancaster Gate—and often went with him on his travels overseas. A letter to Victor in the summer of 1960 ends on a buoyant note:

> We are off to Brussels and later to Zagreb (Friday) Dalmatian Coast (Dubrovnik 4 days) and back by Venice—home on June 28th. The India, Australian, New Guinea, San Francisco, New York trip starts Aug 8.[18]

Stearn and Victor, too, were on good terms at this time, as indicated by the warm tone of their correspondence in 1959/60.

A trip to Johannesburg in 1962 to attend the 50th anniversary celebrations of the South African Chemical Institute is of interest in refuting the view that his aloofness, which some found difficult to accept, could not on occasion be wholly discarded. On this visit he stayed at the home of Mr G. C. C. Gell, who was General Manager of Shell's chemical business in South, Central, and East Africa. They viewed his arrival with some trepidation, fearing that conversation might be difficult, and aware of his reputation for having 'a somewhat difficult temperament and that at times he could be rather gruff and aggressive.'[19] In the event he proved a charming house guest, and this proved the start of a friendship which was renewed when the Gells were posted back to Shell Centre in London in 1963. The last thing he wanted to talk about was chemistry, though he did some simple chemical experiments in the kitchen for the benefit of the Gells' young children. He displayed a catholic interest, ranging from literature to the racial problems of South Africa, from sputniks to politics. He had a seemingly endless fund of amusing stories, and often quoted Hilaire Belloc. They came to recognize him as an essentially simple person, quite unperturbed when confronted with the minor inconveniences of life. They remembered particularly, as so many do, his casual mode of dress:

> Nothing betrayed his simplicity more than his one small suitcase and his unchanging simple form of dress. He wore the same somewhat tatty suit and battered old trilby hat on all occasions. His only concession to convention was when he put on a dinner jacket and black tie for the Chemical Institute's celebration dinner. Even when we went down 3000 feet from the relative cool of Johannesburg into the heat of the Kruger Game Reserve he did not change his attire or remove his tie. Only when he climbed Table Mountain in the Cape—on foot at his own request—did he vary his dress by donning an old pair of grey flannels—a lucky switch for his suit, for he tore the seat of the old flannels on the descent. For his age, around 70 at the time, he was very fit and active, and exhausted my staff in the Cape by the speed at which he tackled the ascent of Table Mountain.

In 1960, Marion married Mark Way at the parish church in Great Missenden, the Dean of Westminster officiating. His appointment as Bishop of Masasi ended in 1959, and he was subsequently appointed to the living of Averham with Kelham, near Newark, which he held until 1971. Although a bishop must have been almost the last person Robert can have expected to have acquired as a son-in-law, it must have been a great satisfaction to see her thus settled. The correspondence with Victor, quoted earlier, is a reminder that he was always concerned for her well-being. A later cause for satisfaction was that Michael was transferred from the home at Brighton to Glen Frith Hospital near Leicester, where—apart from other occasions—Marion was able to visit him on journeys between County Durham and Oxford or London. Robert himself continued to visit him there as long as he was able. (Michael died in 1988.) Marion and Mark had no children, but they adopted two—John and Tonia—in 1960. Thus with Stephanie's three children—Sharon, his namesake

Robert, and Cathie, of all of whom he was very fond—he had in his latter years a pleasing family of grandchildren.

Despite this distressing interlude in his private life, Robinson's activity in other fields was scarcely diminished, regardless of his advancing years. Stephanie—who was extremely fond of him, and adopted him as 'Papa'—was much impressed by his youthful attitude, even though she first knew him when he was in his seventies:

> . . . the biggest things about him which I feel was the sign of genius was his refusal to look back. For him there never was such a thing as 'the good old days', he wanted to always be moving forward and found the world with it's many changes very exciting . . . I don't think he ever realized he was growing older and his mental set never did. During the years that I knew him he was more in tune with what was going on and more excited about the future than most of my contemporaries.[20]

This readiness to embark on new enterprises late in life is nowhere better exemplified than his involvement in scientific publishing. As a contributor, of course, he had had a lifelong involvement, for the prime goal of every academic research worker is to see the results of his work published, and published before any similar results achieved by rivals in the same field. This not only ensures his personal priority, but provides the basis for further general progress in the field concerned. Communication is literally the life-blood of science. Traditionally, the publication of journals, as distinct from books, had been in the hands of learned societies: nearly all Robinson's own research results, for example, appeared in the *Journal of the Chemical Society*. In the late 1930s however, Robinson conceived the idea of a new kind of journal for papers on organic chemistry: this would be similar to the German *Annalen der Chemie*, which published not merely research results *per se*, but memoirs of wider scope. Harold King, then head of the chemistry division of the National Institute for Medical Research, was also interested, but their efforts were frustrated by the outbreak of war in 1939. Afterwards attempts were made to persuade the Royal Society to support something along these lines, but nothing came of them. There the matter rested until the mid-1950s, when Robinson was approached by a commercial publisher, Robert Maxwell, who had become aware of his interest, about the possibility of founding an international journal of organic chemistry. To many of his colleagues, conventional to a degree, the idea of usurping the virtual monopoly of the learned societies would have been—and in the event was—anathema. Apart from other considerations, their journals were a valuable source of income. But with Robinson, with his lifelong connection with industry and commerce, it struck a responsive note.

Circumstances favourable to the achievement of this end began to develop in 1946, when the British government, as part of its programme of post-war reconstruction, decided to encourage commercial scientific publishing in Britain to fill the gap left by the temporary eclipse of Springer of Berlin. Butterworths, a long-established publishing house with interests mainly in the law, were invited to collaborate in founding a joint company with Springer. To promote the scheme Dr Paul Rosbaud, Springer's senior editor, had been brought to Britain immediately after the war, and Butterworth-Springer Ltd was eventually set up. It was a curious partnership, for the Butterworth directors acknowledged that 'they were novices in the world of science'. However, some very eminent scientists agreed to join a Scientific Advisory Panel, and over the next five years about half a dozen specialist scientific journals were launched. However, it seemed to Butterworth that Springer were getting the better of the bargain, and they decided to withdraw. In 1951 they sold their interests to Robert Maxwell, who continued publication under the Pergamon Press imprint, shortly acquiring also the services in an advisory capacity of Rosbaud, who had a 25 per cent interest. From this small beginning—the purchase price was a mere £13 000—Maxwell built up a vast international scientific publishing company, currently publishing more than 400 journals, and with many other interests in the field of book publishing and scientific information recording and retrieval.[21]

In 1956 Rosbaud approached Robinson about the possibility of launching an *International Journal of Organic Chemistry*, and was well received. With characteristic energy and enthusiasm Maxwell followed this up personally by letter and telephone. Finally—at a meeting between Robinson, Maxwell, and Rosbaud on 23 July 1956—agreement was reached in principle concerning the general format of the new journal and Robinson's role in it. It was to be truly international, and Robinson was to be Chairman of an Editorial Advisory board, on terms to be agreed; Dr W. Klyne, Reader in Biochemistry in London University, was to be Editor, and Rosbaud Consulting Editor.[22] The original Advisory Board included many outstanding names, including Robinson's old friend David Ginsburg, the Nobel Laureate Paul Karrer, Carl Djerassi, and Jiri Ruzicka. Nevertheless he had some rebuffs, for the Chemical Society was opposed to the whole idea, and some leading organic chemists refused his invitation to join the Board. The latter was much strengthened, however, when the brilliant American chemist R. B. Woodward—some thirty years Robinson's junior, and generally regarded as his successor as leader in the natural products field—was persuaded to join as co-chairman. So, too, later did Sir Derek Barton, then of Imperial College, whom Robinson had not originally approached, in the mistaken belief that he was hostile. Thus was born *Tetrahedron*, realising Robinson's hopes of twenty years. In his Foreword he declared:

> With the publication of this first part of Volume I a significant and unique enterprise has been launched. The special character claimed for Tetrahedron is its fully international scope, since the new journal is envisaged as a forum for the presentation of original memoirs on organic chemistry contributed from all parts of the world.

Nevertheless, there were teething problems. Some organic chemists still flatly refused to be identified with the new journal, and it was difficult to secure sufficient papers of the right calibre. Robinson was uneasily conscious of this, for he continued his Foreword by saying:

> Without disparagement of the papers herein it must be pointed out that the first number of a new journal must perforce rely on a limited number of enthusiastic authors and the publishers must be content to accept that which lies immediately to hand.

He had certainly not spared himself in trying to make the first issue a success. He contributed no less than five papers himself, Ginsburg rallied with three, and Alexander Nesmeyanov, then President of the Academy of the USSR, two. Such cause as there was for unease lay in other contributions, which came from Belgium, France, Hungary, Israel, Japan, the UK, and the USSR. Refereeing was made more rigorous, standards and speed of publication improved; though diehard opponents remained, they were replaced by younger men with broader vision. Within ten years its circulation exceeded that of the *Journal of Organic Chemistry*, published by the American Chemical Society. Today, its international reputation is well established.

So early in his scientific publishing career, Maxwell must have been well pleased to have secured the collaboration of a man of Robinson's international stature, and he hastened to further the relationship. In March 1956 he had been invited by the Academy of Sciences to visit Moscow and meet leading Russian scientists. This persuaded him that the West underestimated the strength of Russian research in many fields, and he embarked on a five-year programme to publish English translations of Russian scientific books and journals. To further this, he set up the Pergamon Institute, which gained strong support from the US government. In October 1956 he discussed this project with Robinson, and formally invited him to be the first President of the new Institute, to which he agreed.

From this flurry of letters, telephone conversations, and meetings quickly grew a close friendship. In the summer of 1957 Robinson agreed to become Ian Maxwell's godfather, and both he and Stearn came to the christening. All concerned travelled widely and opportunities had to be contrived for meetings:

> I can't tell you how happy Betty and the children enjoyed having you and Sylvia—we were only sorry that it was for such a short time. I am just back from a visit to Stockholm and Paris before going on a trip to America and perhaps will have a chance of seeing something of you and Sylvia.[23]

Maxwell and Stearn immediately found a mutual interest in films. Maxwell's seemingly boundless energy was then finding an outlet in yet another way. He had formed Harmony Films Ltd in 1954 to film stage productions of opera and ballet, including *Don Giovanni*, with Wilhelm Furtwangler as conductor, and Ulanova in *Giselle* with the Bolshoi Ballet. Stearn was then looking for a backer for a series of cartoon films entitled 'Dodo, the Kid from Outer Space'—a sort of early version of ET. It involved a Professor Fingers, who had constantly to be rescued from terrible troubles; Compy, a bird hatched by a computer; and two other characters called How and Why. Maxwell helped to finance it, and eventually some fifty-two films were made, some as Italian versions. The animation was done in Yugoslavia under Stearn's close supervision. There were various profitable spin-offs, such as Dodo dolls, reminiscent of today's Cabbage Patch dolls. All this involved much travel in the interests of promotion as well as production. In 1965 Stearn and Maxwell were featured—with space-age dolls—in the *Johannesburg Star* with Joseph Levine, the President of Embassy Films, who had the distribution rights. The caption claimed that she was flying to New York to effect a contract worth $5 million. Whether this figure is correct or not, it was certainly a highly profitable venture: in the USA the films were very frequently shown by NBC. Robinson, in a note in his unpublished memoirs, was clearly quite proud of Stearn's success:

> These relatively short incidents are bright and witty and in my opinion rank with the best work of Walt Disney.[24]

Again looking back from old age, he genuinely admired her success with a book entitled 'The Dreamer's Dictionary', which she wrote with Tom Corbett. It follows the general line of such books, and to date it has sold over a million copies in fourteen languages. He refers to its 'polished literary style' and the wide acclaim it had received. One wholly unexpected entry in the book surely shows his quirkish sense of humour. It reads:

> *Nobel Prize*. If you dreamed of winning this prestigious distinction, you are being cautioned against arrogance and reminded of what goes before a fall, but a dream of rejoicing in this achievement by a friend or relation is a forerunner to happy family news. Of course if you happen to have actually collected one of these coveted awards, the dream has no significance.[25]

From these beginnings Robinson's involvement with Pergamon increased steadily. In September 1958 Maxwell is inviting his views on *Talanta*, a sister journal to *Tetrahedron*, devoted to analytical chemistry. This was started on the initiative of Professor Ronald Belcher, of Birmingham, who was dissatisfied with existing learned society journals. Its title (Talanta, Greek 'balance') illustrates Pergamon's predilection for classical nomenclature. Mercifully, the initial title *Stoicheion* (Greek 'element') was abandoned because none of those concerned could spell it consistently! It was an immediate success. In March 1959 *Tetrahedron* was joined by *Tetrahedron Letters*. This, too, was an immediate success, both in terms of the number of communications submitted and international subscriptions. Robinson took a very active interest, refereeing many of the articles himself. As time passed, and his 80th year approached, his colleagues on the Executive Board of Editors—especially Sir Derek Barton—became uneasy, feeling that he was setting too low a standard of acceptance, especially in respect of some of his former students. To restore the situation, Barton for several years agreed to share the refereeing with Robinson, unobtrusively ensuring that he himself dealt with the most important communications.[26]

The friendship between the two men prospered, too. Early in 1959, while Stearn was in America before the matrimonial storm broke, Robinson invited Maxwell to the Royal Society Dining Club. Later, to try to relax, it was planned that the two Roberts and Marion should spend a holiday walking in the Alps. At the very last moment, however, Maxwell had to cry off, as he had become involved in the complicated process of being adopted as Labour candidate for Buckingham in the forthcoming general election. He failed to get in then, but subsequently represented the constituency 1964–70. In later years he did find time to spend with the Robinsons in Switzerland—though Stearn was no climber. Early in 1960 the Pergamon Institute, which had been running at a loss, was taken over by Pergamon International Corporation; but Maxwell invited Robinson to continue as a member of a new Advisory council.[27] This was an honorary appointment, but a few months later he came up with a very much more substantial and attractive proposition—the Chairmanship of Pergamon Press Ltd, at an attractive honorarium. Although Robinson accepted at the time, he did not in the end, for a variety of reasons, take up the appointment. Also, in 1960 Maxwell launched another initiative—the Commonwealth Library of Science, Technology, Engineering, and Liberal Studies: with characteristic exuberance he announced that it would consist of 1000 titles—a total finally exceeded—and in 1962 he set off on a world tour to promote it. Robinson was appointed Chairman of the Commonwealth Library, again with an appropriate honorarium: Stearn was to be paid, in New York, a retainer for advice on public relations. Additionally, he became a member of the Board of Editors of the Chemistry Division of the Library. There were also other minor involvements—for example, as section editor of a massive *International Encyclopaedia of Pharmacology* in 1965.

Thus in the course of the years, at an age when most men were relaxing and reducing their commitments, Robinson became closely involved in the activities of what was to become the world's largest publisher of scientific journals. In 1969, however, Maxwell suffered a calamitous setback, when Pergamon Press was taken over by Leasco, an American corporation, and Maxwell lost control. The story of his five-year battle to regain it does not concern us here, and has in any case been recounted in detail elsewhere.[28] It is worth recording, however, that a major factor in his success was the loyalty of the distinguished team of editors he had recruited, many with Robinson's assistance. Leasco were proved sadly wrong in dismissing highly qualified scientific editors as 'two a penny': they were, in fact, the mainspring of the business.

Finally, in the context of scientific journals, mention must be made of one other journal with which he was closely associated for many years. This was *Endeavour*, an international scientific review founded by ICI in 1942: its purpose was to make known the achievements of British science by free distribution throughout the still accessible world. The Royal Society, under its President Sir Henry Dale, gave active encouragement, and several Fellows, including Robinson, became members of an editorial advisory panel. After the war, it assumed a new role. Scientific publication, especially in Europe, was severely disorganized, and there was a need for a truly international review of the progress of science. Accordingly, *Endeavour* policy was reorganized, and it was distributed worldwide in English, French, German, Italian, Spanish, and (briefly) Russian editions. At its peak, total circulation was around 45 000. Robinson continued as a consultant until 1956, when he retired at the age of seventy. During this time—and informally for many years afterwards—he was immensely helpful in suggesting topics and authors, and giving introductions to leading scientists around the world. As Editor, I shall always be very grateful to him. In 1975, ICI decided that it could no longer afford to sponsor the journal, but made it known that—subject to satisfactory safeguards about maintaining the high standard of content and presentation—it would make it over to a commercial publisher. As a very much interested party, I was delighted when Maxwell responded—with characteristic speed and enthusiasm—to my suggestion that Pergamon should take it over.

Sadly, Robinson was then no longer alive to see the journal in which he had taken so much interest brought into the Pergamon fold. Nevertheless, as late as the spring of 1974 he was still interested in major publishing ventures for Pergamon. At that time it was a massive comprehensive work on organic chemistry, and, with a touching faith in my editorial powers, he approached me about the possibility of acting as editor. Reluctantly, for my expertise was not with highly detailed works of this kind, and my chemistry was far too out of date, I went with him to discuss it with Maxwell, then newly back in the saddle at Pergamon—at his London flat. Despite his eighty-seven years and his then almost total blindness, Robinson argued his case extraordinarily lucidly. His remarkable memory was unimpaired. Predictably, however, nothing came of the matter, as it was as clear to Maxwell as it was to me that I would not in any event be a suitable choice. This massive work eventually saw the light of day as the six-volume *Comprehensive Organic Chemistry* edited by Derek Barton and W. D. Ollis.

This long and busy connection with Maxwell and his Pergamon Press has been recounted without interruption for the sake of continuity, but of course it represents only part of his post-retirement activities. His connection with Shell led him to develop a theory—by no means without opponents—of the origin of petroleum, and he wrote a number of articles on this and lectured about it around the world. The travel was not without adventure. Writing to Victor in the autumn of 1962—when he was seventy-six—he mentions casually that he had been involved in a general strike in Brazil and a revolution in Argentina.

Much of the research done at Egham was, of course, directly related to organic chemical problems arising from Shell's growing interest in industrial chemistry. Apart from this, however, he began to venture into what was, for him, entirely new territory, becoming interested in the causes and cure of cancer, especially leukaemia. On the basis of some Canadian research, on the validity of which doubt was later cast, he became convinced that certain relatively simple compounds had a strong prophylactic effect. Many such compounds were synthesized at Egham and tested by Alexander Haddow of the Chester Beatty Research Institute. Shell was not then well-equipped to undertake the concomitant biological research, and collaborated with Ciba, the large Swiss pharmaceutical company, in evaluating the results. Possibly influenced by the nutritional research of H. M. Sinclair, a Fellow of Magdalen, he also came to believe that dietary factors were involved:

There is another embryonic proposal arising from some work by K. Olsen at Copenhagen. He finds that parts of animals (not foetus) contain a substance called 'dispersin' which protects against inoculation of leukaemia cells. It is naturally of tremendous importance to find out what this is. It is of moderately high molecular weight and probably a medium size protein. It seems to be rather stable and may not be destroyed by all forms of cooking. . . . It would be interesting to get statistics of the incidence of leukaemia among meateaters and vegetarians. Many sufferers from leukaemia are children and it could be that such represent the small proportion with deficient dispersin production'.[29]

But dispersin, if it ever existed, proved elusive, and it is significant that he published nothing, despite having given a great deal of time to it. Renée Jaeger, his senior colleague at Egham, inclines to the belief that there was never any substance in the theory, and that Robinson was led astray by unfamiliarity with the complexity of biological systems. If so, he would be by no means the first physical scientist to be so misled.

During his years with Shell he published nearly a hundred papers. The majority of these, however, were on natural product chemistry, and clearly a hangover from his university days. Many indeed were written in collaboration with old associates at the D.P.—Goldsworth, Chakravarti, Muriel Tomlinson. As has been noted earlier, the long list ends in 1974 where it began in 1906, with a paper on brazilin. At the very end of his life he claimed that he had several hundred papers still unwritten. Perhaps this need not be a matter for great regret, however, for many would inevitably have been no more than variations on a theme: by then the whole pattern of organic chemistry, and the interests of a new generation of chemists, had changed radically.

12
The darkening years

Sadly, Robinson's later years were clouded by failing sight, and all the constraints this imposes even when other faculties are largely unimpaired. In his case the affliction took the form of tunnel vision, in which the victim sees the world as it were through a tube of slowly diminishing bore. Its onset and development are insidious: it is not clear when the symptoms first appeared, but a letter to Victor of 21 May 1970 shows signs of real trouble. Its main purpose was to apologize for having forgotten Victor's birthday on the 19th, 'but *tempus fugit* at an alarming rate these days'. The writing is firm, but his sight is clearly troubling him: there is a very wide margin on the left-hand side, and some of the writing trails off over the edge of the page, so that a few words are incomplete or lost altogether. The letter indicates physical hazards also. On the previous day, walking back from the Festival Hall to Shell Centre, he had fallen down three flights of steps and 'made two attempts to destroy a cement wall with my head'. The penalty was luckily no worse than two pigeon-egg lumps on his head and a stiff knee, which he ascribes to 'instinctive protection learned for alpine pursuits'.

In October 1971 he mentions in the course of a letter to Chakravarti that 'My own work has been slowed down by the trouble with my sight': his nephew George Walsh says that in fact he was by that time 'virtually blind'. He was then eighty-four years of age, but notwithstanding he is still busily organized. That evening he was taking the chair at a Royal Society Club dinner, where his guest would be B. Nagy, Professor of Geochronology at Tucson University, Arizona. This reflects partly the fact that he had lectured there years previously, but no doubt more particularly the fact that his inquiries into the origin of petroleum had led to a keen interest in dating the strata of the earth's crust. He plans to go to Manchester quite soon and will then visit Victor at Chesterfield: the following spring he will be going yet again to the USA. Only two years previously he had been there giving lectures on the origin of petroleum, at four centres as far apart as Connecticut and San Francisco. At the latter he indulged his interest in biogenesis by visiting the Ames Space Research Centre, where some research was being done on the creation of simple organic compounds—such as amino acids and peptides—under the influence of radiation.

In these declining years Robinson developed yet another interest—the writing of books. In a way it is not surprising, for it is a sedentary occupation and one demanding a good deal of leisure: thus it suited his new circumstances. On the other hand, the kind of books he had in mind demanded much reference to original written sources, which his blindness made impossible: he had, therefore, to rely on his own prodigious memory, which was not infallible, and the assistance of others under his verbal guidance. Predictably, the results were not wholly successful.

Characteristically, he embarked on not one but three books almost simultaneously. Of these, two need only passing mention in the present context. His lifelong interest in chess led to the publication in 1973 of his *Art and Science of Chess*, written with R. B. Edwards. In 1975 Edwards wrote an obituary notice for *The British Chess Magazine*, and what he said underlines the fact that the habitual remoteness that many attributed to him was often softened:

> We were introduced by a mutual acquaintance and I approached the prospect of working with this formidable man with some trepidation, particularly when it became apparent that his manuscript would need considerable alteration and that certain areas would need complete rewriting. However I need not have feared as I always found Sir Robert easy to approach and prepared to listen to constructive criticism. I regard our collaboration as one of the most rewarding experiences of my life. Whenever I wrote some pages of my own I would forward them to Sir Robert, who would then go through them with a fine tooth comb. I was astonished that a man of his advanced years and poor eyesight should spend so much time in checking every detail. He had a great love for the English language, and was most meticulous on grammar and the correct use of words.

This was followed (posthumously) in 1975 by *An Introduction to Organic Chemistry*, written in collaboration with E. D. Morgan, who had been with him at Shell 1959–66 and then gone on to the University of Keele. It was to have been the first volume of a multi-volume series to be published by Pergamon, but as they were in some difficulties when it was finished it eventually appeared under the imprint of Hutchinsons. In the end only about four of these books ever appeared. In his last years Morgan used to lunch with him at the Athenaeum, and try to tell him what was going on in the world of organic chemistry. He tried also to help him with the organization of a third book which he was struggling to complete.[1]

This third book is of great importance in the present context—though much less so in a broader sense—as it is an autobiography. It is, therefore, clearly the testament by which he wished to be remembered; but, sadly, it is seriously flawed and open to adverse criticism on many grounds. His memory would be ill served if it were taken at its face value. It was planned to be in two parts: the first covering the period up to the time of his appointment at Oxford, the second the remaining part of his life. In the event, only the first part was published, and that posthumously in 1976. Perhaps it was as well that he did not live to see it in print, for even the kindest critics could not gloss over its manifest defects. Almost up to the day of his death he was collecting material for the second part, and much of this is preserved in the archives of the Royal Society in London. It is not, however, anything approaching a complete draft.

Before looking at the provenance of this ill-starred venture its somewhat enigmatic main title, *Memoirs of a Minor Prophet*, needs to be explained. When Holger Erdtman was working in Stockholm in 1932 he went to consult with his professor, Hans von Euler, about the suitability of his applying to do research with Robinson in Oxford. To this Euler replied: 'But isn't he one of the minor prophets?'. This duly came to Robinson's notice, and he recalled it nearly forty years later as a suitable title for his autobiography.

Just when Robinson began to give serious thought to writing an autobiography is uncertain, but Maxwell suggested the possibility to him in October 1965. By his own account,[2] however, several publishers had asked him for such a book, though in the event it was Elsevier who undertook it. It may well be that the Pergamon proposal lapsed because Maxwell was temporarily out of the saddle at the critical time. From the beginning he conceived of it as a scientific rather than general biography; that is to say, it would deal not only with aspects of his academic and industrial career, but with his major achievements in the field of organic chemistry.

His general intentions are clear from letters he began to send to his collaborators at this time. One to Ginsburg in Israel is typical:

I think I told you that I am writing a scientific autobiography and for this purpose I would be extremely grateful to you for a photograph of your goodself . . . Obviously I would wish to include an account of your synthesis of morphine and when you get back I would very much appreciate it if you would kindly send me the most appropriate offprint for this purpose.[3]

Inevitably, it fell between two stools. The general reader would be deterred by the heavy interjections of organic chemistry: dissertations on 'The Fission of some Methoxylated Benzophenones' or 'Structural Relation of Catechin to Typical Anthocyanidins', complete with esoteric formulae, are decidedly less gripping than, say, the account of music-hall life in Manchester before the First World War, or mountaineering in Tasmania in the company of a fellow climber carrying a parrot in a cage because 'he could'nt bear to leave him at home'. Not all readers interested in his life in a general way would agree that 'the description "hetero-enoid" is self-explanatory'. The appeal even to chemists was limited. By the 1970s the simple classical methods on which his success had been built were looking decidely old-fashioned: new techniques abounded, and chemical laboratories were crammed with sophisticated apparatus, much of it computerized. And even among chemists the divisions of interest were such that only a minority, specializing in structural organic chemistry, could follow his arguments. Many of them would look on his accounts as of little more than historical value.

Apart from this, the organization of the book can only be described as chaotic. Accounts of university life in his various appointments are interrupted by long biographies of contemporary chemists and details of complex chemical syntheses and degradations. Although the chapters are ostensibly arranged chronologically—Liverpool (1915–1920), etc.—the chronology is often confusing. At one point we know just where we are on the time-scale, but a few sentences later the context shows that he is describing events relating to quite a different period, though when this may be is difficult to discern. The lack of an index adds to the difficulty of finding one's bearings. Points once noted are hard to relocate, because they might occur almost anywhere.

But the book's worst fault, as a historical source, is that it abounds in minor errors and inconsistencies too numerous to list: for example, a portrait of W. N. Haworth (sometimes misspelt 'Howarth') is described as one of E. L. Hirst; the heuristic method of teaching appears as theoristic; his house, Fairholme in Huddersfield, he calls Fairfield, and so on. Such faults are, however, entirely excusable, for they are inherent in the impossible circumstances in which the work was done. When he started he was in his eighties, when even his remarkable memory had become fallible, and was totally blind. Of necessity the book had to be dictated, and when points of detail had to be checked this had to be done by assistants, who, however willing, could not always find what he wanted or always understand its full implications. At Shell in London he received much sympathetic help from his secretary, Stella Corridon, and at Great Missenden from Rachel Eastwood. Neither was a chemist, and both understandably found great difficulty in coping with the complex chemical nomenclature and formulae with which the text abounds. It is not easy for a lay person to appreciate the subtle but important differences between, say, strychnine and strychnidine, or harmine and harmaline. That the chemical content of the text is as good as it is owes much to the arduous and devoted work of Renée Jaeger. But she acknowledges that from the confused nature of the material with which she was confronted it was a virtually hopeless task from the outset. Nevertheless, working against time, she did what she could to organize it and make it intelligible, particularly as regards the structural formulae and references to the chemical literature.

While it would be misleading not to draw attention to the book's palpable weaknesses, especially as a historical source, this is by no means the same as saying that it is without interest, for it is very revealing about Robinson himself. It shows, for example, that in late life he retained much of the agility of mind which in earlier years had made his lectures so fascinating to some and such a mystery to others: in these he was apt to develop one theme and then dart off in pursuit of another, which had seemingly come into his mind as he was talking. It was not a particularly good recipe for a lecture, and worse for a literary work. Nevertheless, it was the epitome of his style. We see, too, that he had not lost that sense of humour which could be so endearing: the text is enlivened with amusing, inconsequential anecdotes from his younger days. These flashes suggest that if he had given less space to the chemistry and more to the events going on around him, he might have produced a very readable book.

All this poses the question: Why did he embark on this work at all under such severe difficulties? Even in his blindness there were interests—such as chess and music—which he could still actively pursue, and if he were minded to write there were still—on his own estimate—scores of chemical papers for the journals of the learned societies. He was certainly not motivated by the strong feeling of family pride which prompted Philip Robinson to compile the massive *Robinsons of Chesterfield*, for after the first ten-page chapter he says surprisingly little of his family in general, and of Gertrude and Marion in particular. The latter is, in fact, referred to in only a single paragraph, though a draft note in the material accumulated for the unpublished second volume shows that she was to be reintroduced to the reader again, though again only briefly.

Nor can simple vanity and concern for the judgement of posterity have been a major factor. With his distinguished career he could be certain of wide posthumous recognition, in three ways in particular. First, of course, there would be the immediate obituary in *The Times*, carefully compiled beforehand and merely adapted when required. Second, there would be—largely for his peers—an extended notice in the *Biographical Memoirs of Fellows of the Royal Society*. Who the authors would be he could not know, but he could be certain that they would be very familiar with him and his work—partly on the basis of an *aide-mémoire* deposited by himself with the Society—and would do their task very conscientiously. In the event, the authors were Lord Todd and Sir John Cornforth, whose admirable memoir extended to over a hundred pages. Finally, he would undoubtedly appear eventually in the great *Dictionary of National Biography*, decennial (now quinquennial) repository of the lives of Britain's folk heroes from all walks of life: this entry was in the event prepared by Lord Todd.

We must suppose, therefore, that he wanted to make some personal contribution. In part, this may have been to record a synoptic account of the development of organic chemistry during some seventy years of active participation: for this he was certainly uniquely qualified. Again, it has been argued that a major factor was a desire to have a final crack at Ingold, who died in 1970, and against whom he was still inveighing only a few days before his death. He does in fact devote a forty-page chapter to the development of his own electronic theory of organic reaction, with the promise of more to come in the second volume. It includes verbatim quotes from their correspondence, though in one instance, he found, to his chagrin, that a key page in one of Ingold's letters to him had been mislaid. But to use an autobiography as a vehicle for continuing a feud at a highly technical level is very curious. Had he so wished, his reputation was still such that he could have found a publisher willing to publish a book on this subject alone: even later, as we shall see, he found one for an equally specialized and non-topical subject.

Whatever his motivation may have been is now necessarily a matter for speculation: indeed, he may never have rationalized the idea even to himself. What is certain, however, is that the project meant a great deal to him. Renée Jaeger[4] recalls that when she handed him the edited manuscript only three days before he died, he received it very emotionally, literally with tears in his eyes. One cannot help being reminded of the way in which Copernicus received a copy of his *De Revolutionibus* on his deathbed. Robinson deeply appreciated her contribution. In the Preface he stated: 'I wish to express my deepest gratitude for help which vitalised my project.'

The text of the first part was published in 1976, and Robinson's unqualified statement in the preliminary pages that 'Volume II is concerned with the period at Oxford University 1930–1955 as Waynflete Professor of Chemistry, as well as other activities during that period and after retirement' led many to suppose that the work would shortly be completed. In fact, this never happened. Although he left behind a mass of material—now lodged with the Royal Society in London—intended to form the basis of the second volume, it was too incomplete and disorganized to amount even to a rough draft. During 1975 Stearn tried hard to make something of it, but even with the promise of some help from Robert's former colleagues this proved impossible.

One of those whom she approached was Sir Edward Abraham, asking him for help in preparing for press the section on penicillin: 'checking the chemistry and putting in formulae in Robert's MS'. He agreed readily, though not without some private reservations. In 1971 Robinson had been in touch with both Abraham and Wilson Baker, explaining that he was writing a scientific autobiography, and asking them to clarify some particular points of interest concerning the chemical research on penicillin—then already nearly thirty years in the past. This led to a flurry of correspondence between Abraham, Baker, and Cornforth, who were all uneasy that Robinson—in his old age and with failing sight—might write from memory some account which was not strictly accurate. At the same time they had no wish to get involved in any argument about the attribution of priorities. Abraham, writing to Baker, voiced the feeling of them all:

I would be extremely reluctant to get involved in disputes about all this work which was done so long ago, particularly since I have no faith that anything good would come of it. Nevertheless, I should be most grateful to know whether you yourself think that there is anything I should do.[5]

Again, a week later, he is still dubious about the best course to follow:

Robinson made it clear in his letter to me that he did not expect me to comment on a number of his statements, but I will think further about whether I should write to him again. It is, of course, possible that the scientific autobiography will never be ready for publication.

Perhaps the most helpful thing would be to get Robinson to send all of us a final typescript of the relevant parts for comment, but he might well regard this as an impertinence.[6]

As we know, the part of the autobiography containing the reference to penicillin did not appear, so the anxiety was needless. A considerable amount of time, and the exercise of some tact, had gone to no purpose.

That the reconciliation that followed the early matrimonial dispute was genuine and lasting was shown in these dark years when Robert was completely dependent on Stearn. As he acknowledges in the Preface, she gave him much-needed help and advice as he struggled against odds to complete his autobiography. But more than this, as many friends recall, she took pains with his daily comfort. The last time I saw him was at lunch at Grimms Hill Lodge only a few months before he died, and I remember very clearly the patience with which she arranged the food on his plate to make it easy for him to feed himself—'Your meat is at 3 o'clock, your potatoes at 6 o'clock'; and so on.

Even then, he was busy with yet another book. Since before the First World War he had been interested in the alkaloid strychnine, and the chemistry of this substance and its derivatives had lifelong fascination for him. Dispute over the question of priority for elucidating the true structure of strychnine in the 1940s involved him in controversy with R. B. Woodward and Vladimir Prelog. Robinson was anxious to put the record straight, as he conceived it, and also to secure recognition for the contribution of Chakravarti, who had been his co-worker in the D.P. at the time. To further this end, he conceived the idea that they should write a joint monograph, and he wrote to him in February 1961:

What I think possibly would be an excellent thing to do would be for you to write a monograph on Brucine and strychnine, perhaps in collaboration with myself, and to take that opportunity to correct the errors . . . Such a monograph could probably be published either by Pergamon or by the Oxford University Press . . . What do you think of this project?[7]

Chakravarti immediately expressed his willingness to collaborate, but the project languished, kept alive only by a desultory correspondence. Robinson writes again in March 1964:

I think it would be highly desirable to publish this monograph in order to put the whole thing into proper focus and especially, if I may say so, your own fundamental contribution.

However, it proved less easy than he had expected to find a publisher: as Chakravarti himself ultimately realized, the subject was no longer topical, several accounts had already been published, and the methods used had been displaced by much more sophisticated ones. To the end of his life, however, Robinson never accepted this argument. Writing to Chakravarti as late as February 1973 he says:

I wonder whether chemistry would have been richer or the poorer if Baeyer had been able to put indigo into a machine and get the structure right away; we should probably know nothing about the isatins and numerous important synthetic reactions, and so on. It would have been a serious loss, and that is what will happen in the future if the disciples of the physical methods of end-all and be-all in organic chemistry get their way'.

But Baeyer's work on indigo had been done in 1880/3, even before Robinson was born, and no amount of wishful thinking could alter the fact that the chemical world had changed enormously in the mean time.

In turn the Oxford University Press, Wiley, even Pergamon, turned the proposal down. Nevertheless, he persisted—perhaps simply because he hated the idea of being thwarted—and ultimately was successful, concluding an agreement with Academic Press in November 1973. Still no progress was made, however, for Robinson was busy with the second volume of his Memoirs, and seeking Chakravarti's help with the strychnine section of them. Nevertheless, his conscience was clearly troubling him; writing a year later he says:

All this later strychnine work gives me a bad feeling that I have shamefully neglected it, from the publication point of view. Looking back, I cannot really quite understand why that was—perhaps tiredness, but I do hope that something can now be done, though belatedly, to make a partial recovery. Of course, there were the notes in 'Nature', but on looking into the subject again I find that they are rather inadequate to give an idea of the large amount of work involved.

Perhaps realizing that his own contribution must be small, he proposed at the beginning of 1975 that Chakravarti's wife Debi—who would in any case have helped—should be formally added to the publishing agreement as co-author. Sadly, it was all too late. A final letter, written from Shell Centre and dated 6 February 1975, went into the new arrangement in some detail, but was never signed. His secretary, Angela Kingsbury, stamped his signature on it, and sent it to Chakravarti with a covering note saying that he had died on 8 February 1975. He was then in his 89th year, alert to the end.

The judgment of his peers was succinctly expressed by Lord Todd:

Robinson was a man of great mental and physical toughness, with a brilliant and restless mind; he could also be a charming companion with an astonishing memory and interests spreading far beyond his science. Within his science he displayed a penetrating insight which seemed capable of going to the heart of any chemical problem almost instantaneously. His first response to events was usually emotional and he was often impatient of those holding views contrary to his own; these characteristics find a reflection in his scientific work and help to explain its tremendous scope and variety and how it came about that Robinson's name is associated much more with new reactions and theories than with extended syntheses of individual substances.[8]

Marion was with him on the day before he died, and her recollection[9] of their conversation makes a moving and fitting end to this story of his life:

Like so many of us in the Robinson family my father was reputed to be persistent and determined to a degree which could be perverse and obstinate. As a tribe we tend to be independent and resistant to help from others, but inevitably he was physically totally dependent in the days before he died and he seemed content. In his helplessness many notable and lovable qualities came through as I talked to him on the day before he died. He was blind and 88 years old but had continued to dictate chemical reminiscences up to that time. He discussed chemical theories with me, as always mistakenly thinking I understood the nuances, but mainly his thoughts turned to family and friends, particularly the children. He had always been particularly active in his judgement of children's interests, found little gifts to suit them and was at ease in their company.

In that last conversation the breadth of his interests also came through. We talked about mountains and alpine flowers, the gardens he had created and particularly about useful modern shrubs. We exchanged stories and jokes, and he recited some of his favourite pieces from the Ingoldsby Legends and the Breitman Ballads. On that last day he even sang an excerpt from 'Iolanthe'. He was a talented photographer and we looked at some of the enlargements he had developed himself. We also talked about plays and actors and a host of other subjects. He was a life-long Liberal in attitude but we didn't talk at all about the state of this world—nor its possible relation to any other!

Every family has its ups and downs and perhaps the general feeling that I had defected when I decided to work in Tanganyika damaged relationships for a long time, but this was forgotten and we were very happy to be with each other.

A note on principal sources and abbreviations used in the notes

As the preface and text indicate, many sources have been drawn upon in writing the various chapters of this book. The following, however, are of particular importance as being generally relevant.

1. Catalogue of the Papers and Correspondence of Sir Robert Robinson O.M., F.R.S. (1886–1975) compiled by Jeannine Alton and Julia Latham-Jackson of the Contemporary Scientific Archives Centre, Oxford, and deposited in the Library of the Royal Society, 1983 (designated 'RS RR' in the Notes). Most of this material was received from Lady (Stearn) Robinson in 1975 and, after her death on 1 September 1976, from her lawyers. Nearly all the rest come from his daughter Marion, in 1982, and from J. C. Smith of the Dyson Perrins Laboratory. It is a haphazard collection, and the compilers (assisted by Sir John Cornforth) are to be congratulated on having organized it so well. Nevertheless, there remains a small residue of unattributed material. As is to be expected, much of it relates to his research: theses and reports, lectures and papers, notes and notebooks, reprints, etc. In the present context, the most interesting items are the background material, and corrected proofs, of the first (and only published) volume of his autobiography, and the material accumulated for the second volume. So far as letters are concerned, one basic problem is that although he was a prolific correspondent he much preferred to write in longhand, so that there were no copies. What he himself said has, therefore, often to be deduced from recipients' replies, to the limited extent that these survive.

2. Archival material in the possession of Robinson and Sons Limited, Chesterfield. Apart from a considerable number of memorabilia, etc., this includes a large file of documentary material including many press cuttings. Of particular interest are a number of letters to his brother Victor, to whom he was very close. (This source is designated 'Robinson Archives' in the Notes).

3. Robert Robinson, *Memoirs of a Minor Prophet: 70 Years of Organic Chemistry*, **Vol. 1**, Elsevier, 1976. This is, of course, of

fundamental importance, as being his own testament. Nevertheless, as a historic source it must be used with great care, being written when he was well over eighty and totally blind. For a critical appraisal see Chapter 12. (This source is designated 'MOMP' in the Notes.)

4. Lord Todd and J. W. Cornforth, Robert Robinson (1886–1975), *Biographical Memoirs of Fellows of the Royal Society*, **Vol. 22**, pp. 415–527, 1976. This is an admirable account, by two distinguished chemists who know him and his work well, and based partly on material deposited by Robinson with the Royal Society during his lifetime. As is to be expected the emphasis is on his research and professional career: it includes a comprehensive list of his 718 published papers and 32 patents. It says rather little about his relations with industry or his private life.

5. J. C. Smith, *The Development of Organic Chemistry at Oxford*, Parts I and II: Part II is entitled 'The Robinson Era, 1930–1955'. Printed for private circulation, this contains a great deal of general historical information, together with detailed year-by-year information about publications, both papers and books (designated 'JCS' in the Notes).

6. *Natural Product Reports*, **Vol. 4**, No. 1, 1987. Special Robert Robinson Centenary Issue, based largely on twelve lectures presented at the Annual Chemical Congress of The Royal Society of Chemistry, 1986. Topics include his contributions to alkaloid chemistry (K. W. Bentley); to anthocyanins, brazilin, and related compounds (R. Livingstone); to steroids and synthetic oestrogens (J. W. Cornforth); to the early history of penicillin (E. P. Abraham); to theoretical organic chemistry (M. D. Saltzman and J. Shorter); chemistry in Manchester in Robinson's time (W. Cocker); and the Dyson Perrins Laboratory in Robinson's time (M. L. Tomlinson).

Note: Robinson himself, and certain of his more frequent correspondents, are referred to by initials in the following notes, thus: 'RR': Robert Robinson; 'VOR': his brother Victor; 'IRM': Robert Maxwell. Also similarly, in other relevant correspondence, 'EPA': Sir Edward Abraham; 'WB': Wilson Baker.

Notes

Chapter 1

1. Philip M. Robinson and A. Leslie Spence (eds.). *The Robinson Family of Bolsover and Chesterfield.* Chesterfield, 1937.
2. Philip Robinson, *Supplement to The Robinsons of Chesterfield*, Chesterfield, 1961. For much family and business in formation (mainly 1839–1916) see also Crichton Porteous *Pill Boxes and Bandages*: *A Chesterfield Story*. Chesterfield, n.d.
3. Robert Robinson, *Memoirs of a Minor Prophet: 70 Years of Organic Chemistry* (1976) [hereafter refered to as MOMP].

Chapter 3

1. *The Defence of Excellence in Australian Universities*, University of Adelaide, 1978, p. 13.
2. MOMP p. 81.
3. A. J. Birch, *Journal and Proceedings, Royal Society of New South Wales*, **109,** 151, 197.
4. Sydney Morning Herald, 10 February 1913.
5. W. A. Tilden, *Chemical Discovery* and *Invention in the Twentieth Century*. Routledge, London, 1917.
6. G. L. Fischer, *The University of Sydney 1850–1975: Some History in Pictures to Mark the 125th Year of its Incorporation.* University of Sydney, 1975.
7. David Branagan and Graham Holland (ed.) *Ever Reaping Something New: A Science Centenary*. University of Sydney Science Centenary Committee, 1985.
8. *Hermes*, **19,** 43, (1913).
9. *Hermes*, **17,** 73–103, (1911).
10. University of Liverpool Archives 5/11.
11. Register of Shipping Movements, State Archives: personal communication, Graham Holland.

Chapter 4

1. David Nealy, personal communication, November 1985.
2. University of Liverpool Archives S3899.
3. For a general account of British Dyes see W. J. Reader, *Imperial Chemical Industries: A History*, **Vol. 1**, 258–299, 1970 Oxford University Press, Oxford, 1970.

4. MOMP 107.
5. MOMP 106.
6. MOMP 118.

Chapter 5

1. University of St Andrews records (courtesy of the Principal and Vice-Chancellor, Steven Watson).
2. Dr James Craik (personal communication).
3. G. N. Burkhardt 'The School of Chemistry in the University of Manchester', *Journal of the Royal Institute of Chemistry*, September 1954; 'Arthur Lapworth and Others', compiled for private circulation, 1980.
4. W. Cocker 'Chemistry in Manchester in the Twenties, and Some Personal Recollections'. *Natural Products Reports*, **4,** 67 (1987). (A contribution to a centenary tribute to Sir Robert Robinson, organized by the Royal Society of Chemistry as part of their Annual Congress in Warwick in 1986.)

Chapter 6

1. *Endeavour*, **15,** 92 (1956).
2. R. Robinson, *Endeavour*, **15,** 92 (1956).
3. MOMP 141–2.
4. MOMP 229.
5. By Mrs Chris Badden, Records Officer.
6. R. R. Robinson, *British Chemists*, pp. 353–369. The Chemical Society, London, 1947.
7. Association of University College Chemists Newsletter, 1977, p. 14. See also obituary of R. J. W. Le Fèvre by M. J. Aroney and A. D. Buckingham in *Biographical Memoirs of Fellows of the Royal Sociewty*, **Vol. 34,** 1988, pp. 381–2.
8. C. W. Shoppee, 'C. K. Ingold', in *Biographical Memoirs of Fellows of the Royal Society*, **Vol. 18,** 1972.

Chapter 7

1. MOMP 184.
2. Robinson, R. *Outline of an Electrochemical (Electronic) Theory of the Course of Organic Reactions*, (52 pp.). Institute of Chemistry, London, 1932. This was the text of two lectures given to the Institute of Chemistry. He gave similar lectures in Brussels and Stuttgart in the same year, and these were published in French and German respectively.
3. Idem. 'The Development of Electrochemical Theories of the Course of Reactions of Carbon Compounds.' *Journal of the Chemical Society*, 1288, (1947) (18th Faraday Lecture).

4. Idem. 'Some Intramolecular Electrical Effects on the Course of Chemical Change', *Endeavour*, **13,** 173 (1954).
5. Idem. MOMP. pp. 184–228.
6. Lord Todd and Sir John Cornforth, 'Biographical Memoirs of Fellows of the Royal Society', **Vol. 22** (1976), pp. 465–78.
7. Saltzmann, M. D. 'The Development of Sir Robert Robinson's Contributions to Theoretical Organic Chemistry.' *Natural Product Reports*, **4,** 53, 1987.
8. Shorter, J. 'Electronic Theories of Organic Chemistry: Robinson and Ingold.' *Natural Product Report*, **4**, 61 (1987).
9. MOMP p. 17.
10. MOMP p. 25.
11. For a definitive history see C. A. Russell, *The History of Valency*, 1971.
12. MOMP p. 71.
13. MOMP p. 184.
14. *Journal of the Chemical Society*, **111,** 962 (1917).
15. R. Robinson and A. Lapworth, *Nature*, *Lond.* **129,** 278: **130,** 273, (1932).
16. *Journal and Proceedings, Royal Society of New South Wales*, **109,** 151, (1976).
17. *Chemistry and Industry Review*, **44** (1926).
18. Letter, Lowry to Robinson, 19 May 1925.
19. *Journal of the Chemical Society*, 401 (1926).
20. MOMP p. 219.
21. *Journal of The Chemical Society*, 1310 (1926).
22. MOMP p. 221.
23. *Journal of The Chemical Society*,**127,** 513 (1925).
24. *Journal of the Society of Dyers and Colourists*, 65 (1934).
25. *Chemical Reviews*, **15,** 225 (1934).
26. C. W. Shoppee in his biography of Ingold in *Biographical Memoirs of Fellows of the Royal Society*, **18,** 356 (1972).
27. Ingold, C. K., *Structure and Mechanism in Organic Chemistry*, Cornell University Press, 1953.
28. MOMP p. 221.

Chapter 8

1. Those interested in the constitution and administrative complexities of the university during the Robinson era should consult the *Report of the Commission of Inquiry*: 2 vols, Clarendon Press, Oxford, 1966.
2. The history of organic chemistry in Oxford has been recorded in great detail in J. C. Smith, *The Development of Organic Chemistry at Oxford*, Parts 1 and 2,. (254 pp.). Privately circulated, 1955. Pt 2 is devoted to the Robinson era. For convenience, this will be referred to as JCS 1955.

3. JCS 1955, pp. 15, 20.
4. See his entry in the *Dictionary of National Biography*.
5. J. P. V. Dacre Balsdon, *Oxford Life*, 1957.
6. See Keith J. Laidler, *Chemical Kinetics and the Oxford College Laboratories*, 1987.
7. JCS 1955, pt II, p. 2.
8. L. E. Sutton, Memorial Address, Magdalen College, 17 May 1975.
9. M. L. Tomlinson. 'The Dyson Perrins Laboratory in Robinson's Time', *Natural Product Reports* **4**, 73 (1987).
10. George Walsh. Personal communication.
11. Mrs T. S. Sampson. Personal communication.
12. JCS 1955, p. 70.
13. Lord Todd and J. W. Cornforth. *Biographical Memoirs of Fellows of the Royal Society*, **Vol. 22,** (1976).
14. *Journal of the Institution of Chemists (Indian)* (Robinson Centenary Issue), **59,** I, 10–28 (1987).
15. RS RR A54.
16. W. J. R. Reader, *Imperial Chemical Industries: A History*, Vol. II, pp. 81–3 (1975).
17. F. Hilton, personal communication, 1985.
18. RS RR A50.
19. RS RR A50.
20. L. E. Sutton, Memorial Address, 17.5.75.
21. JCS 1955 p. 22.

Chapter 9

1. E. P. Abraham and R. Robinson, *Nature*, Lond., **140,** 24 (1937).
2. T. S. Moore and J. C. Philip, *The Chemical Society 1841–1941* (1941).
3. For a general review of the history of chemical warfare see L. F. Haber, *The Poisonous Cloud*, Oxford University Press, Oxford, 1986.
4. RS RR A53.
5. RS RR A53.
6. M. Stacey, personal communication.
7. R. Robinson, *Biographical Memoirs of Fellows of the Royal Society*, **Vol. 5** (1959).
8. MOMP p. 43.
9. See for example Gwyn Macfarlane, *Howard Florey, The Making of a Great Scientist*, Oxford University Press, Oxford, 1979; Idem, *Alexander Fleming, The Man and the Myth*, 1984 Chatto and Windus, London; Trevor I. Williams, *Howard Florey, Penicillin and After*, Oxford University Press, Oxford, 1984; Ronald W. Clark, *The Life of Ernst Chain, Penicillin and Beyond*, Weidenfeld and Nicholson, London, 1985; E. P. Abraham,

obituary of E. B. Chain in *Biographical Memoirs of Fellows of the Royal Society*, **Vol. 29** (1983).
10. See for example E. P. Abraham, 'Sir Robert Robinson and the Early History of Penicillin', *Natural Product Reports*, **4,** 41 (1987).
11. Clarke, H. T., Johnson, J. R., and Robinson, R., *The Chemistry of Penicillin*, Princeton University Press, 1949.
12. E. P. Abraham, *loc. cit.*, p. 43.

Chapter 10

1. There are many histories of the Royal Society. See for example E. N. da C. Andrade, *A Brief History of the Royal Society*, The Royal Society, 1960.
2. MOMP p. 126.
3. RS RR A67.
4. From Alan Hodgkin's obituary of Adrian in *Biographical Memoirs of Fellows of the Royal Society*, **Vol. 25** (1979).
5. Chesterfield Archive.
6. *The Royal Society at Carlton House Terrace*. The Royal Society, 1967.
7. RS RR A67.
8. *Les Prix Nobel en 1947*, pp. 108–128. Stockholm, 1949.
9. RS RR A67.
10. Philip Robinson, Supplement to *The Robinsons of Chesterfield*, 1961, pp. 106–10.
11. RS RR A67.
12. JCS II p. 14.
13. Letter from RR to David Ginsburg, Weizmann Institute, 28.6.54.
14. *Journal of the Chemical Society*, 2267, (1954).
15. *Nature*, **173,** 566 (1954).

Chapter 11

1. RS RR A52.
2. JCS 1955 II 28.
3. Oxford University Archives File.
4. W. J. Reader, *Imperial Chemical Industries: A History*, Vol. II, pp. 162–182 (1975).
5. W. J. Reader, *op. cit.* pp. 391–410.
6. R. R., personal communication.
7. G. I. Fray, personal communication.
8. RR to VOR 7.1.57.
9. Helmut Meyer, personal communication, 16.3.87.
10. Roald Dahl, personal communcation, 26.8.86.
11. RR to VOR, 9.1.57.
12. RR to VOR, 21/22.3.57.
13. RR to VOR, 7.8.58.

14. RR to VOR, 11.5.59.
15. RR to VOR, 11.5.59.
16. RR to VOR, 14. 5.59.
17. RR to VOR, 17.7.59.
18. RR to VOR, 15.6.60.
19. Mr and Mrs G. C. C. Gell, personal communication.
20. Mrs Stephanie Cottrell, personal communication, 4.12.86.
21. Joe Haines, *Maxwell*, Macdonald, London, 1988.
22. IRM to RR, 23.7.56.
23. IRM to RR, 30.8.57.
24. RS RR A52.
25. Stearn Robinson and Tom Corbett, *The Dreamer's Dictionary: the Complete Guide to Interpreting Dreams* (1974) Panther (Granada), London.
26. D. H. R. Barton, *Robert Maxwell and Pergamon Press*, 1988, p. 431 Pergamon, Oxford.
27. IRM to RR, 20.1.60.
28. Joe Haines, *Maxwell*, 1988. 'Pergamon Lost—Pergamon Regained' pp. 297–334.
29. RR to VOR, 27.3.63.

Chapter 12

1. E. D. Morgan, personal communication.
2. Letter to R. N. Chakravarti 18.2.69.
3. RR to DG 28.10.70.
4. Dr R. H. Jaeger, personal communication.
5. EPA to WB, 6.10.71.
6. EPA to WB, 12.10.71.
7. R. N. Chakravarti, *Journal of the Institute of Chemists (India)*, **59,** 1 (1987).
8. Lord Todd, *Dictionary of National Biography* 1971–1980, Oxford University Press, 1986.
9. RS RR A66.

General Index

Name Index